STRUCTURE–PROPERTY RELATIONSHIPS OF POLYMERIC SOLIDS

POLYMER SCIENCE AND TECHNOLOGY

Recent volumes in the series:

STRUCTURE–PROPERTY RELATIONSHIPS OF POLYMERIC SOLIDS

Edited by

Anne Hiltner

Department of Macromolecular Science
Case Western Reserve University
Cleveland, Ohio

PLENUM PRESS • NEW YORK AND LONDON

Library of Congress Cataloging in Publication Data

American Chemical Society. Meeting (55th: 1981: Atlanta, Ga.)
  Structure—property relationships of polymeric solids.

  (Polymer science and technology; v. 22)
  "Proceedings of the 55th Meeting of the American Chemical Society, held in honor of
the Bordon Award recipient, Professor Eric Baer, during March 1981, in Atlanta,
Georgia"—T.p. verso.
    Includes bibliographical references and index.
    1. Polymers and polymerization—Congresses. I. Hiltner, Anne. II. Baer, Eric. III. Title.
IV. Series.
TA455.P58A39 1981   620.1'92   83-16131

ISBN 978-1-4684-4615-9   ISBN 978-1-4684-4613-5 (eBook)
DOI 10.1007/978-1-4684-4613-5

Proceedings of the 55th Meeting of the American Chemical Society,
held in honor of the Bordon Award recipient, Professor Eric Baer,
during March 1981, in Atlanta, Georgia

©1983 Plenum Press, New York
Softcover reprint of the hardcover 1st edition 1983

A Division of Plenum Publishing Corporation
233 Spring Street, New York, N.Y. 10013

PREFACE

        This book contains a collection of original research papers
which were presented in honor of the Bordon Award recipient,
Professor Eric Baer, on the occasion of the 55th Meeting of the
American Chemical Society (Atlanta, Georgia, March, 1981). The
contributors are present or former colleagues and students who
have worked with him in the Department of Macromolecular Science
at Case Institute of Technology of Case Western Reserve University.
Throughout his work, Eric Baer has attempted to find the relationships of solid state structure and hierarchy to the resultant properties from which specific functions are derived. Although he
has studied many seemingly unrelated subjects, from irreversible deformation, mechanics and yield processes in amorphous polymeric
solids to structural organization and mechanical function of ten-
don, his unique goal has been to develop models from the real
structure that would allow a quantitative description of properties.
Today, this area of "microscience" is rapidly expanding as new and
sophisticated applications of polymeric materials with multifunc-
tional properties are emerging from our understanding and control
of the solid state. The wide-ranging ideas and the original-
ity of Professor Baer's contributions have stimulated many new
concepts which are now widely accepted in the field of high
polymers. The contributions to this volume represent many of the
areas which he has explored.

                                                A. Hiltner

CONTENTS

CRAZING IN THIN FILMS OF "MONODISPERSE" POLYSTYRENE

A. Moreno and E. Baer

Department of Macromolecular Science
Case Institute of Technology
Case Western Reserve University
Cleveland, Ohio 44106

ABSTRACT

A technique to study the relationship between mechanical behavior and craze microstructure of thin films of narrow disperse polystyrene is described. Films about 2-4 μm thick were tested in a conventional tensile machine (Instron). Craze kinetics were observed
through an optical microscope equipped with a motor-driven camera.
In this way, a direct correlation was established between the mechanical behavior and craze growth kinetics. Effects of molecular
weight, thermal history and strain rate on both mechanical behavior
and craze kinetics were studied, and their relationship was elucidated. Statistical analysis revealed that the phenomenon of crazing is
quite complex and depends on several material and experimental variables. The average distance between crazes was found to follow
Poisson distribution, thus confirming the random nature of this phenomenon. The macroscopic strain (elongation) of the material during
the entire deformation history was explained in terms of the number
of crazes per unit area, craze extension and the distribution of
crazes. On the basis of experimental results, a model is suggested
from which the elastic modulus of crazed material can be derived.

INTRODUCTION

Research into various aspects of the phenomenon of crazing has
played an important role in shaping our understanding of the fracture
behavior and the toughening mechanism of polymers. Recent results
offer important clues to the structural nature of the glassy phase
of the macromolecular chain. Since the publication of exclusive

1

reviews by Rabinowitz and Beardmore [1] and Kambour [2], researchers have continued to seek specific knowledge of the relationship of the mechanical behavior to the molecular structure of glassy polymers. A recent publication [3] gives the mechanistic steps associated with craze formation from localized cavitation to fibrillation through a local plastic deformation. The significant role played by molecular weight was clearly identified. Whereas void nucleation was not influenced by molecular weight, the stability of localized plastic deformation and subsequent fibrillation was strongly dependent on molecular weight. Only polymers with molecular weights above the critical entanglement molecular weight [4] were found to exhibit stable craze growth. This molecular weight requirement was subsequently confirmed by Fellers and coworkers [5,6]. In the latter study it was suggested that crazing is greatly diminished in polystyrene with a molecular weight less than twice the critical entanglement. Tensile measurements in polystyrene [7,8] and fracture energy estimation in polymethylmethacrylate [9] invoke similar conclusions. Nevertheless, a systematic study relating craze growth to molecular weight is missing from the published literature.

This paper describes results of a study of the effect of molecular weight on craze growth and mechanical behavior; both were observed simultaneously. The effects of rate of deformation and thermal history are also considered.

EXPERIMENTAL

Materials

Samples of narrow molecular weight distribution atactic polystyrene were used for the tensile experiments. The polymers were supplied by the Pressure Chemical Company; molecular weights and molecular weight distributions are listed in Table 1.

Preliminary experiments indicated that there is very little, if any, difference in the mechanical behavior or the corresponding crazing kinetics between 6.7×10^5 and 1.8×10^6 molecular weight polymers. It was decided therefore, to use 1.0×10^5 and 1.8×10^6 molecular weight polymers to investigate the effect of molecular weight. In the rest of this discussion these two polymers will be referred to as low molecular weight (LMW) and high molecular weight (HMW), respectively. At the end of the discussion, however, results obtained from 6.7×10^5 molecular weight polymers will be briefly reviewed. All of the experimental results were averaged over at least five runs. In one experiment, test results from fifteen identical samples were averaged.

Table 1. Molecular Weight, Molecular Weight Distribution and Glass
          Transition for Polystyrene Samples

Short Form	Mol. Wt.	Mol. Wt. Distr.	Tg (^{o}C)
LMW	1.0×10^5	1.06	99.0
$\sim$	6.7×10^5	1.15	99.8
HMW	1.8×10^6	1.33	99.9

Film Preparation

The films were cast by dipping clean glass slides in solutions
of polystyrene in xylene. The solutions were 5%, 9% and 13% by
weight for the 1.8×10^6, 6.7×10^5 and 1.0×10^5 molecular weight
polymers, respectively.

The solvent was evaporated in two steps: first, in air at room
temperature for about one hour; then overnight in a vacuum oven at
about 87^oC. This temperature is at least 10^oC lower than the glass
transition temperature for the three polymers, as shown in Table 1.
These temperatures were calculated by using an extension of the
Fox-Flory theory [10,11].

After annealing, the samples were either quenched in ice water
or slow cooled (about 10^oC/hr) under vacuum. The films were then
removed from the glass slides by flotation on the surface of a water
bath and dried on filter paper. The thickness of film specimens
prepared in this manner ranged from 2.5 to 3.5 μm, as determined by
infrared spectroscopy [12].

Fig. 1 shows a comparison between the IR spectra of polystyrene
and xylene, the casting solvent. The lack of absorbance in the 1380
cm^{-1} region in the PS spectrum indicates complete solvent removal.
This is further supported by the imperceptibility of the sharp peak
at 800 cm^{-1} in the PS spectrum.

Sample Preparation and Testing

Specimens with a diameter and gauge height of 2 cm were cut with
a specially designed template. Subsequently, samples were annealed
to remove any excessive edge effects caused by the cutting procedure.
The edges of annealed samples appeared smooth at 60x magnification.

A conventional Instron testing machine was used for the tensile
experiment. A 500 g load cell, for which the minimum full scale
range is 10 g, was required to measure the small loads inherent in
thin film work. The tensile stress-strain measurements were carried
out at two different strain rates: 0.1% and 0.025% for the 1.0×10^5

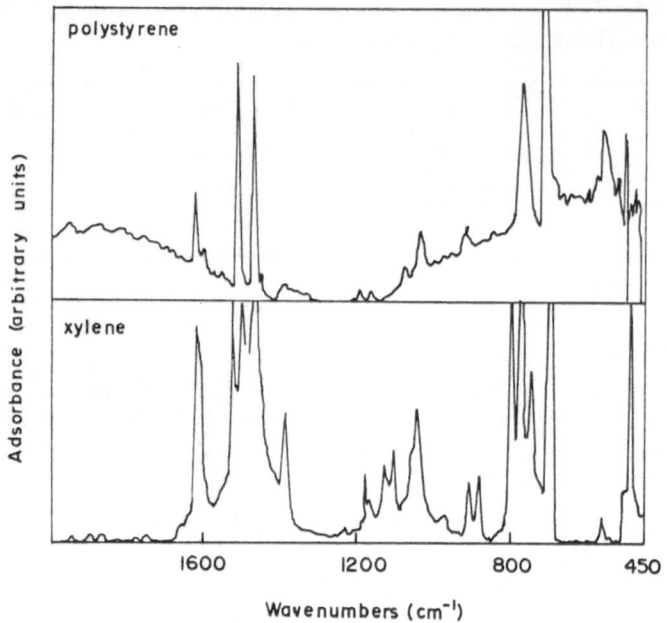

Fig. 1.   IR spectra for a polystyrene film and xylene, from which
the film is cast indicating complete solvent evaporation.

molecular weight PS; and 0.1% for the 6.7 x 10^5 and 1.8 x 10^6
molecular weight PS.

Craze propagation as a function of deformation was observed
through a Nikon optical microscope mounted directly on the Instron,
and was photographed with a motor-driven Nikon FM camera capable of
more than one frame per second. Craze growth was analyzed from
micrographs at 100x magnification, in association with the
corresponding mechanical behavior.

RESULTS

Stress-Strain Measurements

Error Analysis. Because the dimensions of the specimen are
very small, measurements from the Instron machine can have significant errors, even when a very sensitive load cell is used. Errors
were found to be about 50% for the initial modulus, and 20% and 25%
for the initial stress and elongation, respectively. The two latter
values are reduced to 10% at elongations well beyond the elastic
limit. As for measurements of the ultimate elongation, additional
variations might have occurred due to defects in the edge of the
film specimen. Considering the dimensions of our specimens and the

equipment sensitivity, this error seems normal. Other investigators
[13,14] have reported similar or even higher levels of error in
much larger specimens.

Effects of Molecular Weight. Fig. 2 shows the tensile behavior
of LMW and HMW films at a strain rate of 0.1% min^{-1}. Both were
quenched from the annealing temperature ($\sim$90°C) by plunging into ice
water. The LMW film exhibited only elastic (Hookean) behavior, followed by a brittle fracture at a much lower elongation (less than
2%) than the fracture strain in the HMW material. The fracture
stress was also consistently lower. The HMW polymer, on the other
hand, displayed different regions: an elastic region, a plastic
linear region, and a smooth transition zone between the two. The
strain-hardening observed is unusual for polystyrene, and the fracture elongation (4-6%) is much higher than that reported for bulk
specimens.

The stress at fracture for HMW PS was higher than that of LMW
film. A "yield" stress of about 18 MPa for the HMW PS was estimated.

Effect of Strain Rate and Thermal History

Low Molecular Weight. Fig. 3 shows the effect of strain rate
on the mechanical behavior of LMW quenched PS films. At the higher
strain rate (0.1% min^{-1}), the polymer behaved in a Hookean manner
until fracture. The elongation of fracture was about 1% for all
samples. At the lower strain rate (0.025% min^{-1}), the same material
exhibited more ductile behavior with very pronounced instability at

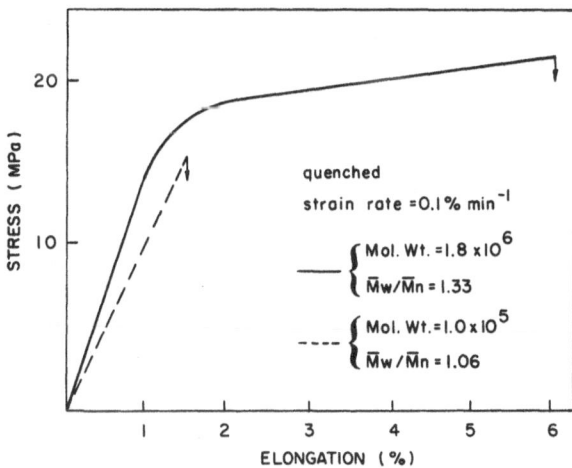

Fig. 2. Effect of molecular weight on the mechanical behavior of
         thin films.

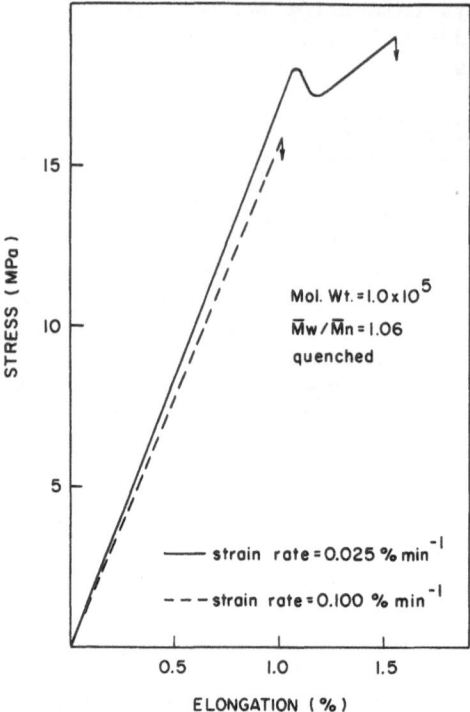

Fig. 3.   Effect of strain rate on the mechanical behavior of low
          molecular weight films.

the "yield" point. The stress dropped at about 1% elongation, then
increased continuously, showing a strain-hardening behavior until
fracture at less than 2% elongation and at a stress of about 20 MPa.

     Similar but more dramatic effects were found due to thermal
history. As illustrated in Fig. 4, when LMW films were deformed at
a strain rate of 0.025% min^{-1}, slow cooled samples showed extraordinarily weak and brittle behavior with unusually low ultimate
elongation (less than 0.7%). Simultaneous visual observations
indicated no crazing at all.

     High Molecular Weight. As shown in Figs. 5 and 6, the effect
of strain rate and thermal history on the mechanical behavior of
HMW PS is not as pronounced as in the case of LMW polymer. A
ductile-like behavior was observed in all samples. An elastic zone,
a plastic zone and a smooth transition "yield" region generally
characterized the tensile behavior of the material at both thermal
histories and both strain rates. Two subtle differences, however,
can be distinguished from the figures. 1) The transition is more
abrupt at the higher strain rate (0.25% min^{-1}), and 2) the transition spreads over a 1% elongation at the lower strain rate (0.1%

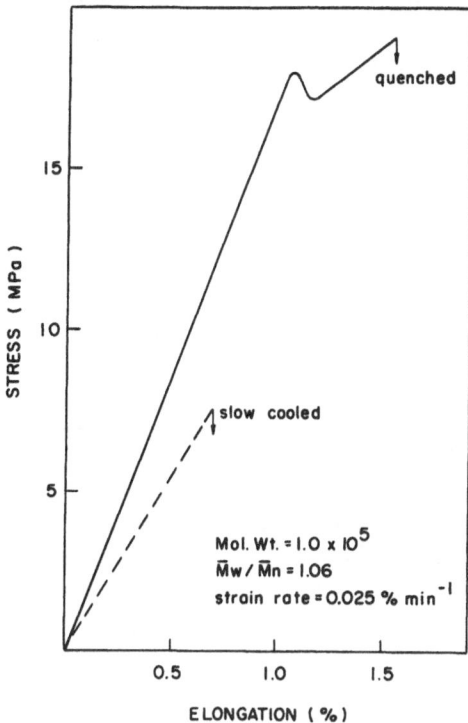

Fig. 4.   Effect of thermal history on the mechanical behavior of
          low molecular weight films.

min^{-1}), with occasional slight "yield" instability. The quenched
samples and those tested at a higher strain rate exhibited lower
stress at all stages of deformation. The latter conditions also
showed a smoother "yield" transition.

     In all HMW samples, quenched or slow cooled, and at both strain
rates applied, profuse crazing occurred within the transition region.

Morphology of Crazing

     Figs. 7a and 7b show optical micrographs of HMW films at 2.6%
and 4.1% elongation, respectively. Fig. 8 shows an optical micrograph of LMW film at 1.7% elongation. It is clear that crazes in
LMW polymer are longer, thicker, and more distant from each other
than crazes in HMW polymer. Crazes in LMW films were always found
to be perpendicular to the direction of the applied stress, and only
a few were so close that they might interact. On the other hand,
crazes in HMW films (Figs. 7a and 7b) were not always found to be
perpendicular to the stress direction; they were more numerous, and
it appears that the majority were interacting, and some even over-
lapped, especially at higher elongations.

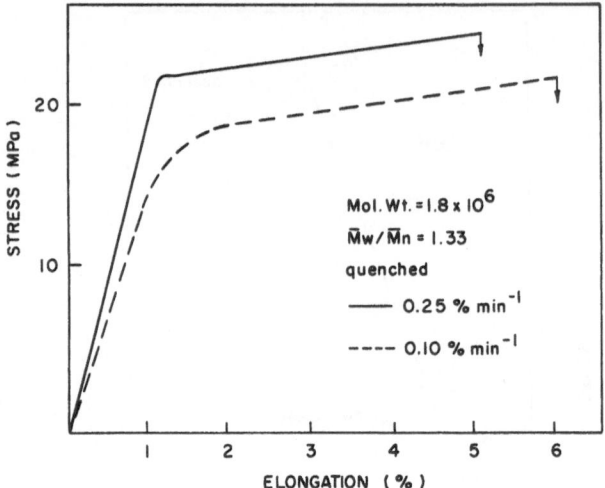

Fig. 5.   Effect of strain rate on the mechanical behavior of low
          molecular weight films.

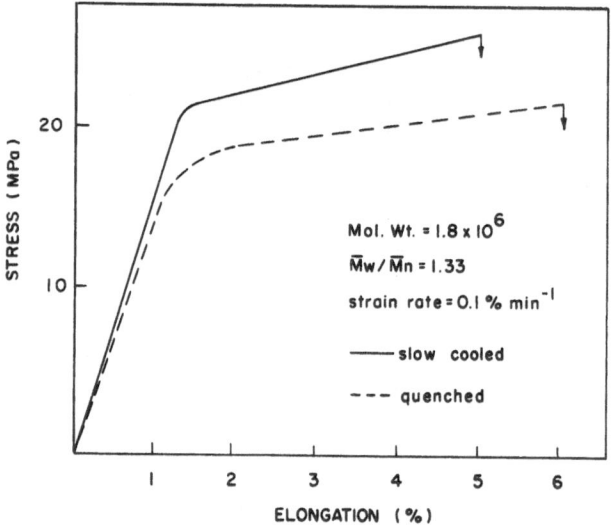

Fig. 6.   Effect of thermal history on the mechanical behavior of
          high molecular weight films.

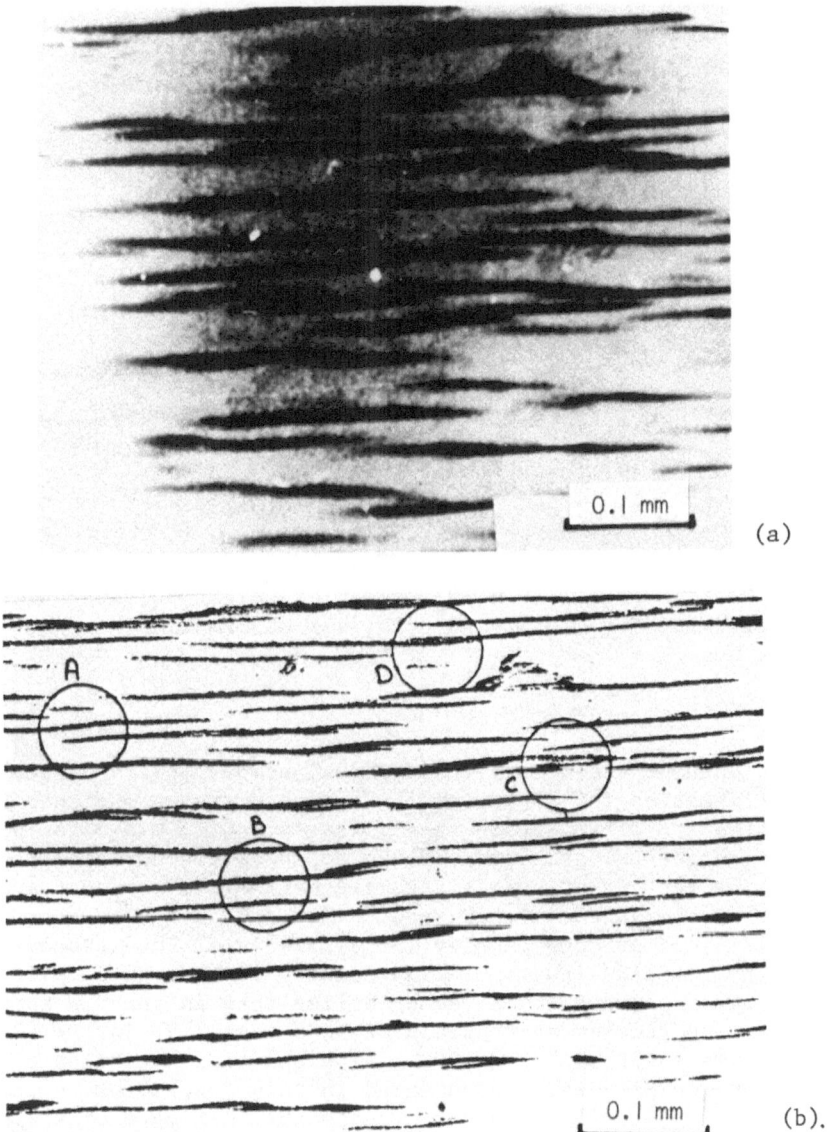

Fig. 7.   (a) Micrograph of 1.8×10^6 molecular weight polystyrene
film strained at 0.1% min^{-1} to 2.6% elongation.   (b) Micrograph of 1.8×10^6 molecular weight polystyrene film
strained at 0.1% min^{-1} (same as in 7a) to 4.1% elongation.

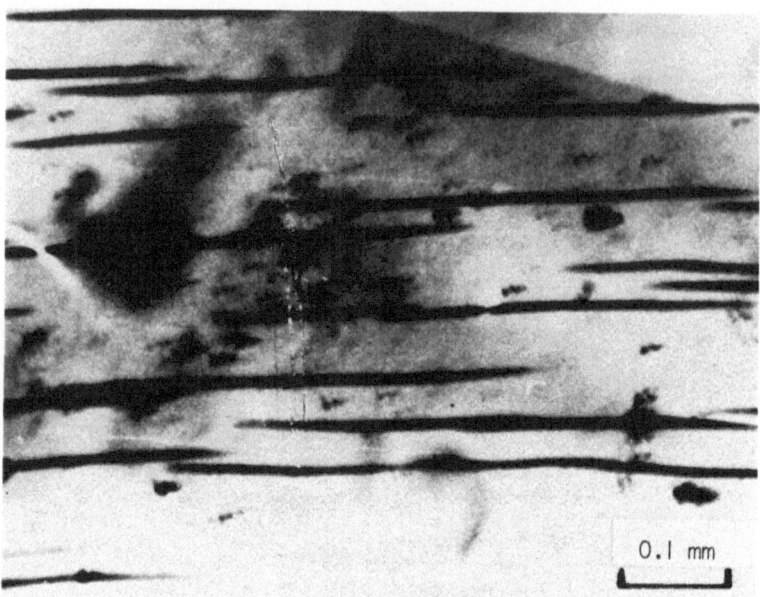

Fig. 8.   Micrograph of 1.0×10^5 molecular weight polystyrene film
          strained at 0.025% min^{-1} to 1.7% deformation.

Kinetics of Craze Growth

     Fig. 9 shows the density of crazing (number of crazes/cm^2)
versus deformation for HMW PS and LMW PS polymers.   Note that the
strain rates are different in the two cases: 0.1% min^{-1} for HMW and
0.025% min^{-1} for LMW samples.   It was not possible to compare both
materials at the higher strain rate since LMW films did not show
crazing when elongated at that rate.   This can easily be seen from
the magnitude of crazing density in this material when tested at
0.025% min^{-1} (solid line).   Crazing density in the LMW samples was
found to reach an asymptotic value earlier than in the HMW samples.
This level was reached at a strain of less than 1.5% in LMW PS,
compared with about 3% for HMW PS.   Although the initiation of the
first craze was not easily documented in this experiment, extrapol-
ation of the two curves provides a reasonable estimate of the strain
(and stress) at which crazes start.   A value of approximately 0.7%
is obtained from the two curves of Fig. 9.   The final value for the
crazing density in the HMW films is more than six times as high as
in the LMW films.

     In Fig. 10 a direct comparison of typical individual craze
growth history in LMW and HMW samples is depicted.   In the LMW
polymer, craze growth is generally a steep linear function of

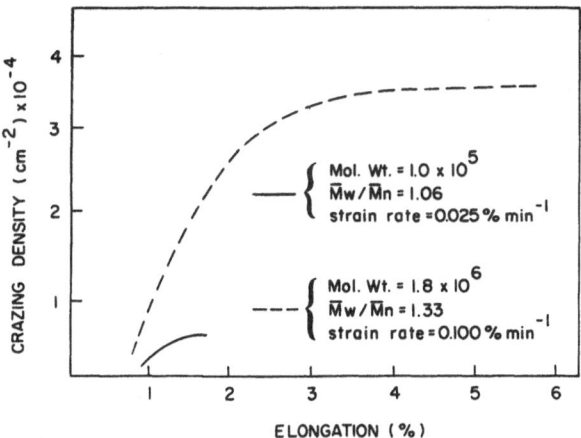

Fig. 9. Crazing density versus elongation for high molecular weight
         (solid line) and low molecular weight (dotted line) films.

elongation throughout craze development. Crazes grew much faster in
the LMW films than in the HMW films, although a few crazes showed
slower growth due to interaction with other crazes. A clear example
of this case is the craze growth indicated by the open squares: at
the point indicated by the arrow, another craze interacted with this
particular craze, and both continued to grow at a much smaller rate.

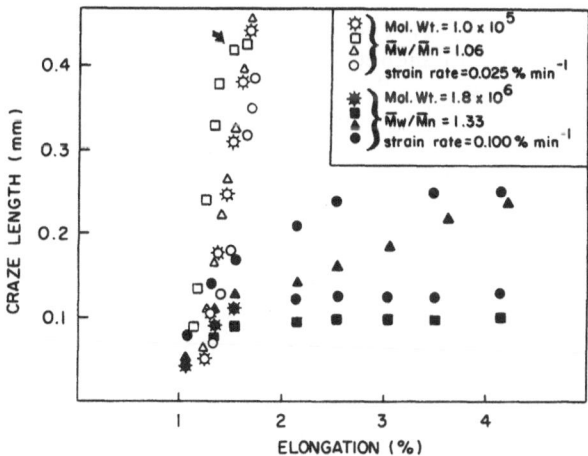

Fig. 10. Craze length of four typical crazes as a function of
          elongation for high molecular weight (solid symbols) and
          low molecular weight (open symbols) polystyrene.

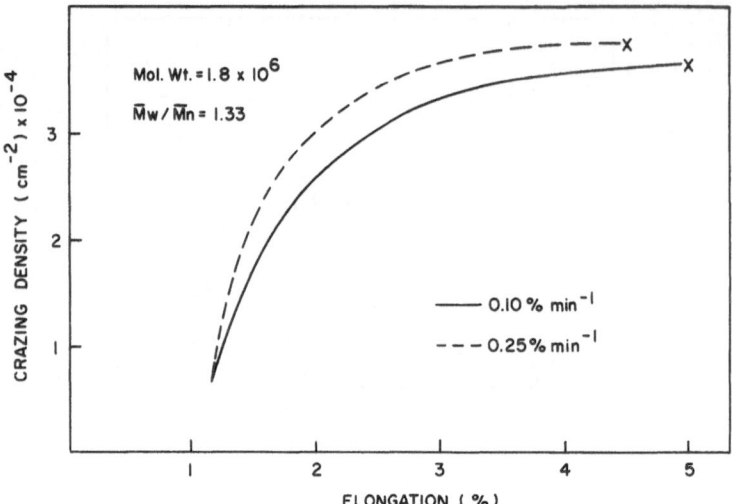

Fig. 11.   Effect of strain rate on crazing density as a function of
           elongation for high molecular weight polystyrene.

Two other significant features of craze growth for HMW PS films are
apparent in Fig. 10.   First, crazes that were initiated at lower
elongation grew faster than crazes initiated at higher elongation.
Second, craze length increased exponentially to asymptotic values,
with the exception of a few isolated crazes, which continued to grow
throughout the entire course of elongation, albeit at a slower rate
(solid triangles).

     Figs. 11 and 12 display the effect of strain rate on the kinet-
ics of craze growth for quenched HMW polymers.   Each curve of Fig.
12 represents an average of five crazes.   The crazing density (Fig.
11) is higher for the sample elongated at higher strain rate (0.25%
min^{-1}) than for the sample elongated at lower strain rate (0.1%
min^{-1}).   This relationship is reversed for craze length (Fig. 12),
i.e., longer crazes occurred at lower strain rate, and vice versa.
In both cases a symptotic value for crazing density and craze length
is approached at about 3% elongation.

     Fig. 13 is a composite plot of mechanical behavior and craze
growth kinetics: stress-strain, crazing density, craze length, and
distance between crazes.   The four curves are averaged over 15 identical samples tested under the same conditions, i.e., quenched HMW PS
films strained at 0.1% min^{-1}.   In all cases, profuse crazing occurred
in the knee region of the stress-strain curve, then an asymptotic
value of crazing density was approached in the plastic region at
about 3% elongation.   By extrapolating the craze density curve, an
elongation of 0.7% is obtained for the appearance of the first craze.

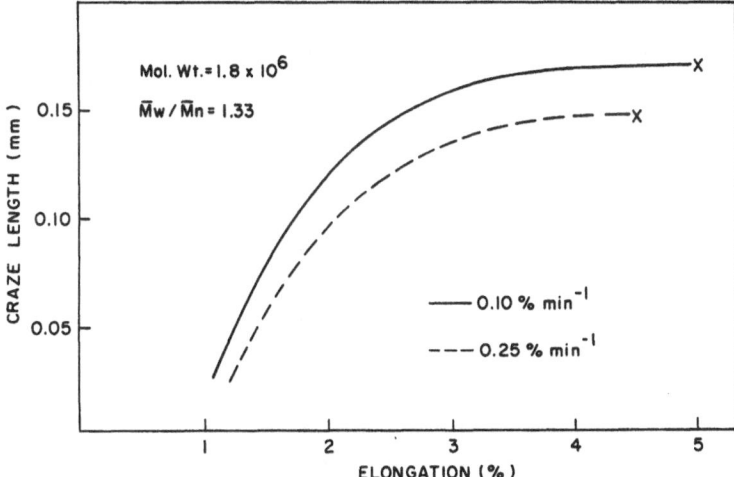

Fig. 12. Effect of strain rate on craze growth as a function of
          elongation for high molecular weight polystyrene.

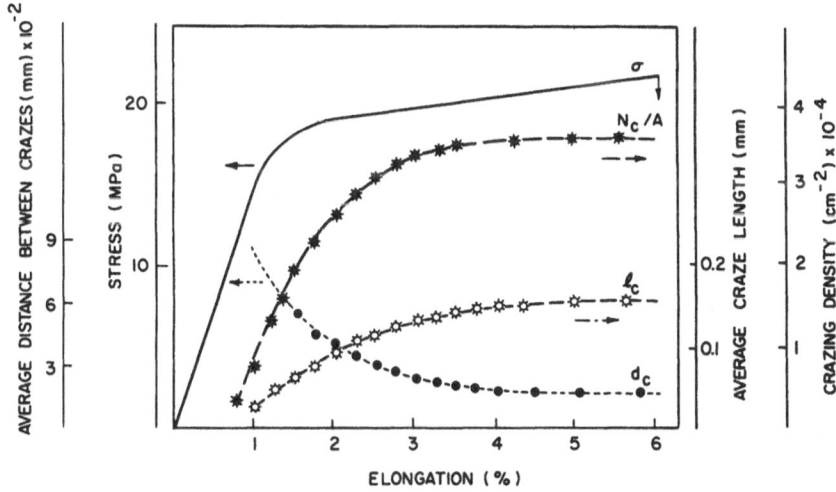

Fig. 13. Relationship of crazing density history (N_c/A), craze
          length (l_c) and distance between crazes (d_c) to the
          mechanical behavior of 1.8×10^6 molecular weight poly-
          styrene strained at 0.1% min^{-1}. All plots are averaged
          over 15 identical samples.

Both the craze length and the distance between crazes are averaged
over 40 crazes in each of the 15 samples tested. As the elongation
was increased, the craze length increased slowly and approached an
asymptotic value of about 0.15 m at about 3% elongation. The average distance between crazes exhibited decreasing exponential behavior; and again, an asymptotic value (of about 10 μm) was approached
at higher elongations.

    In order to appreciate the randomness of the phenomena investi-
gated, sampling of the typical behavior of individual crazes is
shown in Figs. 14 and 15. The average distance between crazes versus
elongation for nine different samples of the fifteen tested in this
experiment is shown Fig. 14. Considerable fluctuation is observed,
especially at low deformation. At higher deformation, however, all
the curves show the same asymptotic value (around 10 μm). Fig. 15
shows the large fluctuation in crazing density exhibited by twelve
identical samples. The statistical average of this behavior is
shown in Fig. 13.

    Fig. 16 shows the distribution of the distance between crazes
observed in a typical sample at 1.3% elongation, which is well within
the knee region. The histogram exhibits the observed distribution,
whereas the curve shows the calculated Poisson distribution. The
average distance between crazes represents the parameter λ of the
Poisson distribution [15]: $P(x) = (\lambda e^{-\lambda d})/\Gamma(d+1)$ where d is the
distance between crazes, $\Gamma(d+1)$ is the gamma function and λ is the
average distance between crazes. Obviously, the calculated
distribution adequately describes the phenomenon.

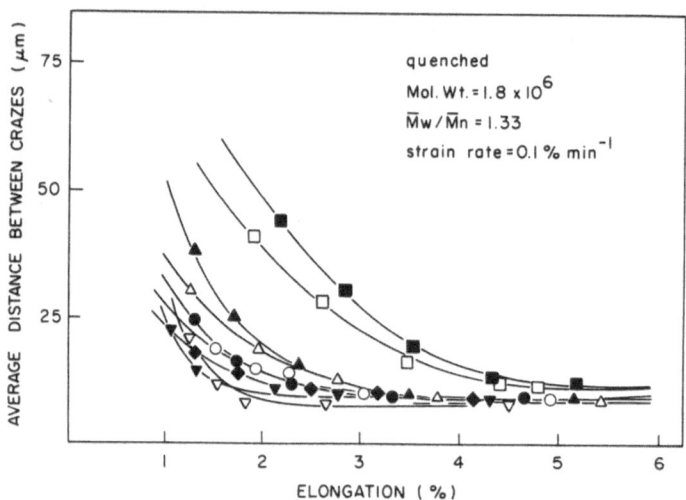

Fig. 14. Average distance between crazes as a function of
          elongation for nine identical polystyrene films.

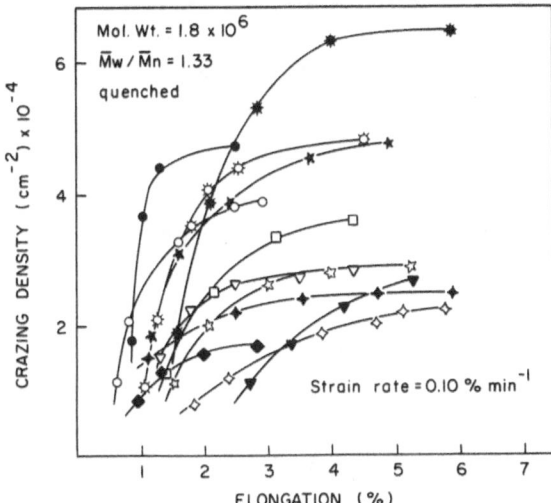

Fig. 15. Crazing density as a function of elongation for identical
samples of polystyrene, showing the statistical nature
of the phenomenon.

DISCUSSION

Stress-Strain Behavior

    It is instructive at the onset of the discussion to recall some
of the well established aspects of craze growth briefly referred to
in the introduction. Microscopic evidence [3] had suggested that
void formation, the craze nucleation step, does not show strong
dependence on the molecular weight. In contrast, stable growth of
these voids into plastic zones, which subsequently coalesce and form
the fibrillar material, shows critical dependence on molecular
weight. In addition, the strength of the craze fibrillar material
was found to increase with increasing molecular weight. This image
conforms well with macroscopic measurements [5], which show that
craze initiation does not depend on molecular weight. In contrast,
craze fracture stress for polystyrene increases with molecular
weight in a way similar to that of tensile strength.

    It is reasonable to invoke the ideas summarized above to explain
major differences in the observed mechanical behavior of HMW and LMW
polystyrene. Evidently, the higher molecular weight polymer is
capable of more stable craze formation. This is clearly shown by
the stress-strain behavior of LMW and HMW films in Fig. 2. LMW film
was much weaker and showed no plastic deformation, although both
materials were deformed at the same rate. Fracture of LMW PS occurred at a stress close to the yield stress of HMW PS. No crazes

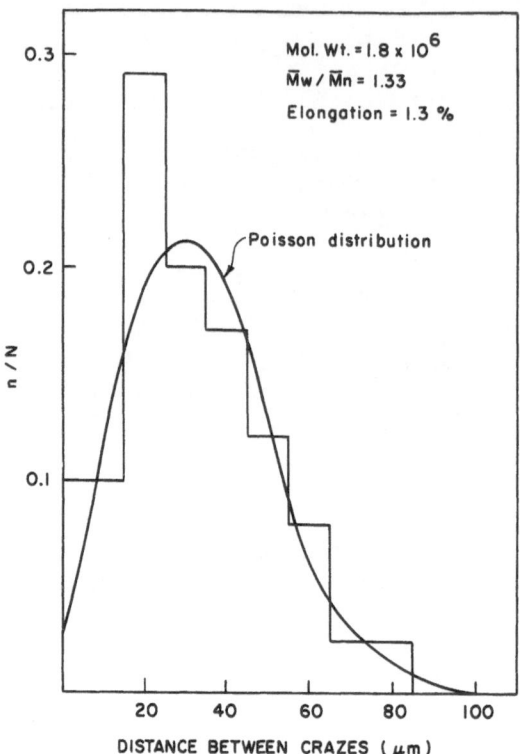

Fig. 16.   Histogram and Poisson distribution of the distance between
           crazes for high molecular weight film elongated at 1.3%.

were visible in LMW films; however, profuse crazing took place in
HMW films.

     Due to the relative abundance of chain ends, and according to
the idea that voids form more easily at such points [16], it is
reasonable to assume that the LMW polymer has a greater tendency for
void formation. At the same time, this polymer is less capable of
stable fibrillation and therefore prone to earlier failure without
visible craze growth.

     At a lower strain rate (Fig. 3), another sequence of events may
take place, resulting in a different behavior. LMW PS was found to
exhibit a more ductile-like behavior, and craze formation could be
observed. Most likely, the time scale dominating the deformation
process at this strain rate allows fiber formation, and consequently,
craze development was facilitated. The observed craze yield instability constitutes yet another interesting feature of the deformation process. In this region, a few crazes form and propagate
rapidly. Fig. 10 suggests that the craze material was not stable

enough to support the applied stress. Therefore, the crazes propagate rapidly until equilibrial stress redistribution is attained.
The relative instability of these crazes (in LMW film) is further
supported by the low fracture strain (and stress) compared with the
HMW samples. The same line of reasoning may be used to explain the
dramatic effect of thermal history on the mechanical behavior of
LMW PS (Fig. 4).

Films cast from HMW PS responded in an essentially different
manner from that of LMW films. This becomes clear by comparing Figs.
3 and 4 with Figs. 5 and 6. Yield instability appeared in HMW PS as
the strain rate was increased (Fig. 5), while in LMW PS, yield instability appeared as the strain rate was decreased (Fig. 3). This difference in behavior appears related to the characteristic relaxation
times for each polymer. The yield instability (Figs. 3 and 5) occurred concurrently with the sudden appearance of a large number of
crazes. Were the LMW material deformed at a strain rate lower than
0.025% min^{-1} (which falls beyond the instrument capability), perhaps
a stable yield transition would have been observed.

Invoking concepts of the theory that treats crazing as a first
order phase transition [17,18], the yield instability can be considered as a superstressed state (corresponding, e.g., to a superheated
state). Changes in molecular weight and processing conditions (thermal history) yield a material structure with specific characteristic
relaxation times. If the characteristic time of deformation (strain
rate) is higher than that of the material, instability is observed
in the stress-strain behavior, accompanied by the sudden appearance
and speedy growth of crazes. Indeed, this was observed in LMW
(Fig. 3) and, to a lesser extent, in HMW material.

Fig. 17a shows hypothesized Gibbs free energy functions for uncrazed and crazed materials. The point 0 represents the transition
from the uncrazed to the crazed phase at a critical stress σ_* (craze
yield stress). This condition should be fulfilled if the deformation
rate matches the "characteristic time" of the polymer. If, on the
other hand, the rate of deformation is higher, the system may exist
in a metastable superstressed state (indicated by 1). That is, craze
initiation will be delayed until a higher stress (σ_*'), at which
point the system drops from the metastable (1) to the stable (2)
state (Fig. 17a). The superstressed state is described here as
similar to other familiar metastable states, e.g., superheated or
supercooled states. The G-σ relationship in Fig. 17a corresponds
to the stress-strain behavior shown in Fig. 17b.

In view of the documented randomness of craze initiation (Fig.
16) and the complexities inherent in their propagation (Fig. 15), it
should be noted that the above discussion represents a crude estimate of the average behavior of many interfering events. Nevertheless, an early analysis of direct measurements of the stress-strain

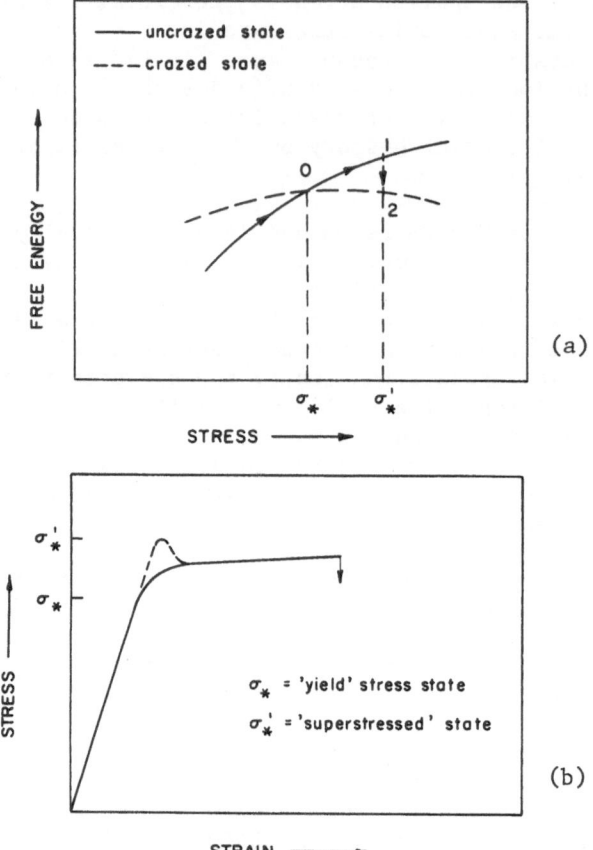

Fig. 17. (a) Hypothetical Gibbs free energy-stress relationship
          for crazed (dotted line) and uncrazed (solid line) materi-
          al. Crazing occurs at 0 under normal (slow loading) or
          at 1 under superstressed conditions (fast loading). (b)
          Schematic stress-strain curves for normal (solid line)
          and superstressed (dotted line) phase transition of a
          material undergoing crazing.

behavior of single crazes by Ulan and Hoffman [19], on the same
film as was used in this experiment, leads to conclusions compatible
with our propositions.

     Higher stress levels in the plastic region of the stress-strain
curves due to the faster deformation rate (Fig. 5) and slow cooling
of HMW PS (Fig. 6) can be elucidated in light of morphological observations recently reported by Murry and Hull [20] and more recently by Trent, Palley and Baer [21]. Crazes with more dense fibrillar
structure were found to grow at higher strain rates, a phenomenon

which perhaps occurs in slow cooled films as well. These crazes
probably resist deformation better than crazes with less fibrillary
density and therefore exhibit the higher plastic deformation stress
observed.

It appears that slow cooling reduces the characteristic size of
intrinsic heterogeneities and thereby increases craze initiation
stress (Fig. 6). An equivalent effect can be caused by faster deformation, i.e., fewer heterogeneity sites are realized by the applied stress at a higher strain rate. These suggested events may
explain the rise in yield stress due to higher strain rate (Fig. 5)
and slow cooling (Fig. 6).

Plasticity of HMW Films

The unusual plastic deformation displayed by HMW material (Figs.
2, 5 and 6) can be readily understood in light of the results of the
morphological kinetic investigations. Comparison of the micrographs
in Figs. 7a and 7b indicates that plastic deformation from 2.6% (a)
to 4.1% (b) must have occurred by crazed growth and to a lesser extent by crazing density increase. Whereas craze extension can be
directly inferred from these micrographs, craze widening cannot be
inferred. It is evident from Fig. 13, however, that craze widening
must contribute significantly to plastic deformation, particularly
at higher elongations. The fact that average craze length and
craze density become constant at about 3-4% strain suggests craze
widening as the major source of plastic deformation.

Craze Growth Kinetics

The dramatic influence of molecular weight on the kinetics of
craze growth as typified in Fig. 10, is considered here. We have
already pointed out that different strain rates had to be applied

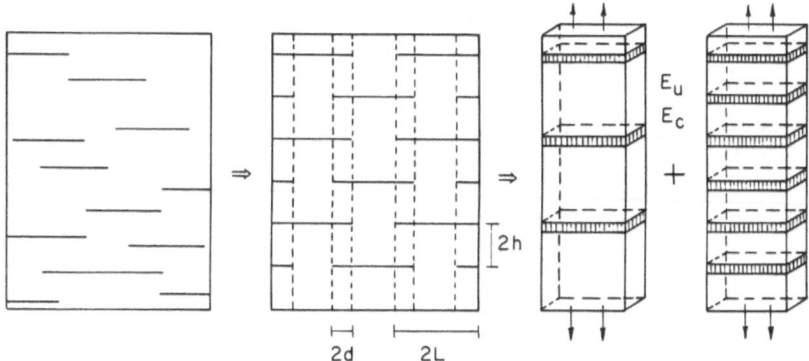

Fig. 18.   Representation of the model used to calculate the elastic
modulus of crazed matter.

in order to compare craze growth kinetics in LMW and HMW polymers.
Initial examination of the data indicated that craze growth in LMW
films was higher than in HMW films. If the difference in strain
rate is taken into account, another relationship evolves.

The total deformation in the sample at any time $\varepsilon(t)$ is the sum
of crazed $\varepsilon_c(t)$ and uncrazed $\varepsilon_u(t)$ deformation, i.e.,

$$\varepsilon(t) = \varepsilon_u(t) + \varepsilon_c(t) \tag{1}$$

By differentiation we obtain:

$$\dot{\varepsilon} = \dot{\varepsilon}_u + \dot{\varepsilon}_c \tag{2}$$

where $\dot{\varepsilon}$ is the strain rate used in the experiment. Then, if we sup-
pose that all post-yield deformation is due only to the deformation
of the crazed matter, $\dot{\varepsilon}_u$ can be neglected. The craze-induced deformation ($\dot{\varepsilon}_c$) can be expressed as the sum of three different terms.
This is based on the fact that the deformation of the crazed matter
is due to craze extension, widening, and to a lesser extent, an
increase in the number of crazes. Therefore,

$$\dot{\varepsilon}_c = (\partial\varepsilon_c/\partial\ell)\,\dot{\ell} + (\partial\varepsilon_c/\partial N)\,\dot{N} + (\partial\varepsilon_c/\partial w)\,\dot{w} \tag{3}$$

where $\dot{\ell}$, $\dot{N}$, and $\dot{w}$ are the rates of craze extension, widening and
density, respectively. When only a relatively small number of crazes
are present (initial part of the curves in Fig. 10), we can neglect
the second term. It is also reasonable to assume that craze widening is too small in this region, and accordingly, we may also neglect
the third term. Therefore, from equations (2) and (3) we obtain

$$\dot{\varepsilon} = (\partial\varepsilon_c/\partial\ell)\,(d\ell/dt) \tag{4}$$

A true estimate of the craze extension rate can be estimated accord-
ingly, since $\dot{\varepsilon}$ is known and $(\Delta\ell/\Delta\varepsilon)$ can be evaluated from Fig. 10.
From this calculation, the following values are obtained:

$$\dot{\ell}(\text{LMW}) = 36 \ \mu m/min \qquad \dot{\ell}(\text{HMW}) = 26 \ \mu m/min$$

Although the craze extension rates obtained here are normalized with
respect to strain rate, it is still obvious that LMW crazes grow
faster than HMW crazes. This point definitely confirms our idea
about the relative instability of LMW crazes compared with HMW
crazes. In turn, this may also explain why LMW materials, in
general, show more brittle behavior relative to HMW PS.

A Model for Calculating the Elastic Modulus of Crazed Material

Combining mechanical measurements with crazed morphology facil-
itates the development of a simple model from which the elastic

modulus of the crazed material (fibrillar and voids together) can
be derived. From the micrographs of Figs. 7a, 7b and 8, a crazed
film appears as a composite of crazed and uncrazed materials alternating in a sandwich structure, which can be approximated by the model
shown in Fig. 18. Crazes are represented by equally long and equally
spaced lines. Due to craze overlap (Figs. 7a, 7b and 8), a two
component system with a less dense component I and a more dense
component II is adopted. To simplify the treatment, both crazed
and uncrazed constituents are assumed to be elastic materials with
different constants. On this basis, an equation can be derived for
the elastic modulus of the crazed material (E_c):

$$E(\epsilon) \cdot (L+d) = (L-d)/((\eta/E_c) + (1-\eta)E_u) \\ + 2d/((\eta/2E_c)^c + (2-\eta)/2E_u) \qquad (5)$$

where η is the volume fraction of the crazed phase, $E(\epsilon)$ is the
modulus of the crazed film, and E_u is the modulus of the initial
uncrazed film.

All quantities except E_c are obtainable from Fig. 13, where
$E(c)$ is taken as secant modulus at the deformation being considered.
At 2% elongation, for example, the secant modulus is $E(0.02)=0.96$
GPa. With an elastic modulus of 1.5 GPa for the uncrazed material,
the modulus of the crazed material (E_c) is found to be 0.28 GPa.
Clearly, the elastic modulus of the crazed phase is less than the
elastic modulus of the uncrazed phase. This point may seem incompatible with the idea that the crazed matter is oriented. However,
the orientation is related to the fibers and not to the overall
crazed phase, which is composed of fibers and voids. If it were
possible to deform a single fiber, the elastic modulus would probably
be higher. A lower elastic modulus for the crazed matter is a
rubber-like phase with an elastic modulus of one-half the elastic
modulus of a glassy polymer [22]. In a recent report [23], an
elastic modulus of about 0.13 GPa was estimated for the craze
fibers, using a fiber volume fraction of about 0.3.

Results for the 6.7×10^5 Molecular Weight Films

Some samples of 6.7×10^5 molecular weight PS were used in the
investigation. The general behavior of these samples is similar to
that of the HMW samples. Yield stress and strain hardening behavior
are much the same. The only noticeable difference is in the elongation at fracture; 6.7×10^5 molecular weight PS exhibits a fracture elongation of 2.5-4%, compared with 6% for HMW material. This
observation seems to agree with the observations of Rudd [24]. He
noted that lower molecular weight samples had weaker plastic zones
than higher molecular weight samples. Earlier we suggested that
plastic deformation above 4% elongation probably occurs by craze
widening. Therefore, it is reasonable to assume that 6.7×10^5
molecular weight films undergo earlier failure because the fibrillar

structure within the craze is not strong enough to sustain the stress needed to incorporate new craze material during the plastic phase. Again, this difference could be related to the entanglement molecular weight M_e. The 6.7×10^5 molecular weight is about 20 M_e, and the HMW (1.8×10^6) PS is more than 50 M_e. The difference is significant enough to consider it as a cause of the different behavior exhibited by the two polymers.

CONCLUSIONS

Our results emphasize the importance of the thin film technique developed here in understanding the phenomenon of crazing and how it relates to the mechanical behavior of the polymer. Effects of molecular weight, strain rate, and thermal history on mechanical behavior have been established by examining samples made from polymers of narrow molecular weight distribution. Below is a summary of these results.

1) High molecular weight polystyrene (1.8×10^6) exhibits more ductile behavior than lower molecular weight polymer (1.0×10^5).

2) By increasing the strain rate or by slow cooling the samples, crazing is suppressed only in the low molecular weight films.

3) In the high molecular weight polystyrene, either by increasing the strain rate, or by slow cooling the samples, the "yield" stress increases and the fracture strain decreases.

4) Craze initiation is always observed at the deviation from linearity of the stress-elongation behavior.

5) The mechanical behavior of the high molecular weight polystyrene exhibits three different phases:
   i) Perfect elastic (no crazes observed).
   ii) Nonlinear behavior (intensive craze formation).
   iii) Constant stress deformation (crazes grow in length and possibly in width).

6) Crazes of low molecular weight films grow faster than those of high molecular weight films, reflecting the relative instability of the low molecular weight material.

7) For high molecular weight material, crazes that are initiated earlier grow faster.

8) Crazes are fewer and longer in low molecular weight samples than in high molecular weight films.

9) Length, density and distance between crazes reach a final constant value after a certain time, and additional deformation seems to be due to the thickening of the crazes.

10) There is a critical distance between crazes (10 m) which indicates a craze "saturation" condition of the system.

11) Craze nucleation exhibits a Poisson distribution, showing the randomness of the crazing phenomenon.

12) The elastic modulus of the crazed matter, estimated from the sandwich model proposed, is about one order of magnitude less than the initial modulus of the uncrazed film.

ACKNOWLEDGMENT

    The authors wish to acknowledge the generous financial support
of this research by the National Science Foundation through the
Material Research Laboratory at Case Institute of Technology under
grant No. DMR80-20245. The authors also wish to thank Professors
Chudnovsky and Moet for their valuable contributions to this work.

REFERENCES

1. S. Rabinowitz and P. Beardmore, "Craze Formation and Fracture
    in Glassy Polymers", Critical Reviews in Macr. Science, E. Baer,
    P.H. Geil and J.L. Koenig (Eds.), Vol. 1, CRC Press (1972).
2. R.P. Kambour, J. Polymer Sci.: Macromol. Rev., 7, 1 (1973).
3. S. Wellinghoff and E. Baer, J. Macromol. Sci.-Phys., B11(3),
    367 (1975).
4. J.D. Ferry, "Viscoelastic Properties of Polymers", 2nd edition,
    Wiley, New York (1970).
5. J.F. Fellers and B.F. Kee, J. Appl. Polymer Sci., 18, 2355
    (1974).
6. J.F. Fellers and D.C. Huang, J. Appl. Polymer Sci., 23, 2315
    (1979).
7. A.N. Gent and A.G. Thomas, J. Polymer Sci., A-2(10), 571 (1972).
8. L. Nicolais and T. DiBenedetto, J. Appl. Polymer Sci., 15, 1585
    (1971).
9. R.P. Kusy and D.T. Turner, Polymer, 17, 161 (1976).
10. D.T. Turner, Polymer, 19, 789 (1978).
11. R.F. Fedors, Polymer, 20, 518 (1979).
12. J.G. Grasselli, Ed., "Atlas of Spectral Data and Physical Con-
    stants for Organic Compounds", CRC Press, A19 (1973).
13. J. Vlachopoulos, N. Hadjis and A.E. Hamielec, Polymer, 19, 115
    (1978).
14. N. Brown and S. Fisher, J. Polymer Sci.: Polymer Phys., 13,
    1315 (1975).
15. H. Cramer, "Mathematical Methods of Statistics", Princeton
    University Press, 203 (1946).
16. T.R. Steger and L.E. Nielsen, J. Polymer Sci.: Polymer Phys.,
    16, 613 (1978).
17. A. Chudnovsky, I. Palley and E. Baer, J. Mater. Sci., 16, 35
    (1981).
18. A. Chudnovsky and E. Baer, Bull. Am. Phys. Soc., 26(3), 464
    (1981).
19. M.J. Ulan and R.W. Hoffman, Bull. Am. Phys. Soc., 26(3), 464
    (1981).
20. J. Murray and D. Hull, J. Polym. Sci., Part A-2, 8, 1521 (1970).
21. J.S. Trent, I. Palley and E. Baer, J. Mater. Sci., 16, 331
    (1981).
22. A.N. Gent, J. Macro. Sci.-Phys., B8, 597 (1973).

23. A. Moet, M. Rackovan and E. Baer, SPE 38th Annual Technical
     Conference (1980) 252.
24. J.F. Rudd, J. Polymer Sci.-Polym. Letters, 1, 1 (1963).

THERMOREVERSIBLE GELATION OF CRYSTALLISABLE POLYMERS

AND ITS RELEVANCE FOR APPLICATIONS

A. Keller

H. H. Wills Physics Laboratory
University of Bristol
Royal Fort
Tyndall Avenue
Bristol BS8 1TL

DEFINITIONS AND SCOPE

By definition macromolecular gels are networks of macromolecules imbibed with a liquid, which is a potential solvent, with chain continuity throughout the whole macroscopic sample. In most traditionally studied gels the junctions are chemical cross-links, and hence permanent. In the gels we shall be concerned with here the junctions are physical associations, and in particular they correspond to crystals, which comprise only a comparatively small portion of the chains. These crystals, like all polymer crystals, form under appropriate supercoolings when solutions of polymers are cooled. On heating, the crystals dissolve and thus the network falls apart, the gelation - dissolution on cooling and heating being repeatable, hence the term "thermoreversible gels".

Such thermoreversible gels have been known for long (e.g. ref. 1). Our particular contribution to the subject rests essentially on identifying the junctions as being crystals, and in the study of the nature of these crystals and their effect on gelation directly by X-ray diffraction aided by other auxilliary methods diagnostic of the crystalline state. In this we were greatly aided by examining the gels in the stretched state where the diffraction effects in question appeared in a most prominent form in addition to providing novel information on the oriented state and on the orientation process in general.

Originally we came across the phenomenon of thermoreversible gelation in our studies of crystallisation as one of the manifestations of the crystallisation process. It soon emerged, however,

25

that the implication of the gelation effects in question were reaching far beyond the scope of our original interest in crystallisation. In fact the ramifications of gelation studies soon got us into a diversity of subjects embracing a large body of polymer science, and even technology, which we are currently pursuing to a greater or lesser extent circumstances permitting.

Table 1 below contains a list of the diverse subject areas with which we have become associated one way or another as a result of our work on thermoreversible gels. Or rather more importantly, the items listed turned out to be subjects to which such gelation is relevant and on which gelation phenomena have thrown new light. The purpose of this article is to lay out this list and place it on the map, in order to support the statement of their connection with thermoreversible gelation and to indicate interrelations such as are apparent in the present rather fluid state of the subject. In aid of this I shall follow the list in the table item by item and shall "tick off", so to speak, each in turn as I have dealt with it, occasionally reverting to some, as and when interrelations between them are becoming apparent.

As already stated our attention was drawn to thermoreversible gels through the study of crystallisation from solution. In the first instance this crystallisation was in the quiescent state. Effects belonging to this category are denoted as Class I in the Table. Later, quite unexpectedly, gelation phenomena of thermoreversible character have turned up in the course of experiments involving agitation and/or flow. This class of effects we term Class II. There are obvious similarities between I and II to be commented on later.

As a further, general statement the studies in question involve a number of different polymers. Not all of them are displaying all of the effects in question with equal clarity, or conversely, some polymers are better suited for a particular investigation than others. Nevertheless, in most cases there is an underlying thread common to all. Even so there are certain effects which, so it appears at present, are specific to a given polymer of a particular chemical constitution. However, even in such latter cases either the effect itself, or the polymer in question as such, is of special interest per se to justify its inclusion into this survey.

CLASS I GELS

1. Crystallisation and Crystal Morphology

In suitable superheatable systems (e.g. isotactic polystyrene, i-PS) two kinds of crystallisation can be observed as the solution is being cooled: (A) the usual platelet formation, and (B) gel forming

Table 1

An enumeration of consequences and implications of the thermoreversible gelation phenomenon. Class I refers to strictly quiescent
systems and Class II where flow or agitation is involved at some
stage of the preparative procedure.

---

(1) Crystallisation
          The crystallisation process
          Morphology

(2) Anomalous orientation effects

(3) Copolymer crystallisation

(4) Steroregularity (PVC)   Class I
          Amount
          Distribution
          Implication for the polymerisation

(5) Plasticisation - a technological issue

(6) Chain conformation of polyolefins
          Conformational variants
          Methodology of conformational analysis

(7) Gelation and fibre formation
          Formation, production and structure
          of fibres
          Properties of fibres - high modulus and
          strength

(8) Flow induced crystallisation   Class II

(9) Origin and nature of flow induced gelation
          Transient and permanent junctions
          Relaxation times

(10) Hydrodynamics of polymer solutions

(11) Issues relating to entanglements

---

crystallisation[2]. The two are clearly distinct and identifiable.
Platelet crystals (A) are particulate and form a turbid suspension,
while the gel (B) by definition sets into a permanent shape; in its
usual form it is transparent to the eye. (A) we identify with the
products of the usual chain folded crystallisation, while (B) with
crystallisation specially designed to create connectedness, where the
crystals are the junction points; the latter we visualise as micellar
crystals (Fig. 1). While clearly intermediate cases, say platelets
with tie molecules so as to form gels, also need to be counted with,
I believe there is a definite distinctness about the classification
expressed here to become apparent below.

     A major feature of (B) is that it sets in only at very high
supercoolings in polymers which are sufficiently supercoolable, and
only at such high supercoolings. While the need for high molecular
weight is clearly an important factor in creating connectedness, this
will only become operative at a sufficiently high supercooling,
otherwise particulate platelet precipitate forms even at identical
concentrations (which are in the range of 1-10%). In a suitable
temperature interval (A) and (B), i.e. platelet and gel, may form
and exist together. In this case the gel is turbid to the eye.*

     We interpret the above effects as follows. Formation of (A) is
to be visualised along the usual channels applicable to chain folded
crystals. (B) in my interpretation would be the consequence of the
fact that the critical nuclei need to be only very small to become
stable at those high supercoolings. Thus only a very few molecular
segments would need to come together in order to "stick", so to
speak, into a stable. entity. Such could arise from the chance
encounter of a few separate molecules thus forming junction points
for an incipient network. With the "micelle" being so small the
cumulative strain at the end surfaces, where the chains emerge (due
to the much discussed and disputed space filling problem arising at
such interfaces[3]), will still be limited and will thus not yet
represent a prohibitive obstacle, as it would in case the micelle
were more extended laterally forming a broad lamella. This point, as
may be recalled, is being invoked in favour of the need of at least a
certain amount of adjacently reentrant chain folding (e.g. ref. 3),
so as to account for the geometric feasibility of extended lamellae.
Folding of this kind will be predominantly intra-molecular which will
thus lead to particulate structures, as opposed to the network forming tendency of the more intermolecular crystallisation implicit in

------

*Concentration clearly played an important part, high concentration
promoted gel formation. Nevertheless we were unable to define a
critical concentration below which gelation ceased as it appeared
to be merely a matter of time for how long we were prepared to
wait.

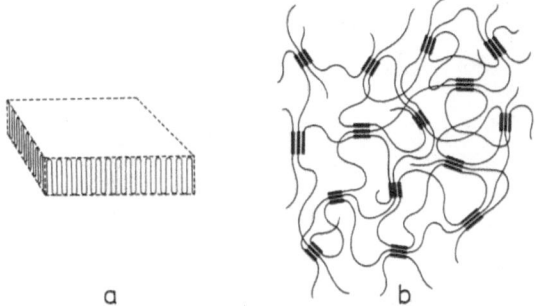

a   b

Figure 1.   Schematic representation of the two basic morphologies
            arising in the course of crystallisation in a quiescent
            system.   a) Lamellae, which in dilute systems are separate
            entities.   Here the molecular arrangement is chain folded
            (drawn as adjacently reentrant and regular for the present
            sketch).   b) Micelles creating an overall connectedness
            which leads to a network where the crystals are the
            junctions.   In solutions this gives rise to a gel which
            is thermoreversible.

the micelle formation. Accordingly, the small lateral size, the
micellar morphology and the network forming tendency should be
causally related. Indeed, the limited lateral size of gel forming
crystals is apparent from X-ray line broadening, the comparison
between lamellar chain folded crystals in this respect, in case
where both coexist, being particularly striking (Fig. 2).

    Thus, as we have seen, the micellar crystals are distinct as
regards conditions of formation and as regards crystal size. To
this needs adding that the lattice parameters in the gel forming
crystals (B) are usually slightly larger, indicating also lower
perfection as compared to the platelet crystals, a feature also apparent from Fig. 2. In addition crystals (B), at least in the present class I, melt at a lower temperature than crystals (A).

    Based on all the above facts, I venture to suggest that we are
dealing not merely with stages in a gradation of crystal development
(although such may also exist) but with the existence of two distinct
crystal morphologies corresponding to the above classification. It
happens, that in solutions this is manifest by the even qualitatively
apparent effect of gelation. If the same occurred in melt crystallisation, i.e. at sufficiently high supercoolings (A) gave room to
(B) type morphology, this would not be recognized by such a primary
test. The reality of the latter suggestion, namely that the same
distinction may exist in highly supercoolable melt crystallised polymers, is strongly indicated by the salient works of Berghmans et al.[4].

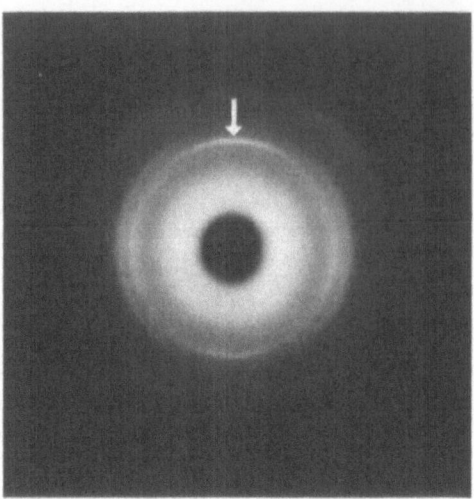

Figure 2.   X-ray diffraction pattern of a stretched PVC (40°C
            suspension polymerized) gel fibre after drying. The
            meridional 5.2Å reflection is marked by arrow[17]. Draw
            direction vertical.

     Further, in principle, a connected crystalline gel network could
also form within an initially compatible two-component system as a
result of a liquid-crystal phase separation of any one or both of the
two components on suitable lowering of the temperature according to
whether one or both of the components are crystallisable. (In the
latter case two interpenetrating gel networks would result.) This
interesting possibility of engineering particular blend morphologies
relying on network forming crystallisation has been exploited by
Wellinghoff, Shaw and Baer[45] in the case of the compatible blend
system of poly(phenylene oxide) (PPO) and isotactic polystyrene (i-PS).

2.   Anomalous Orientation Effects

     The existence of anomalous orientation effects, as manifest by
X-ray diffraction, abound in orientation studies of high polymers, in
crystalline gels in particular. In our experience such effects occur
in a particularly pronounced form when crystals (A) and (B) coexist.
Based on a wide ranging experimental material the following systematic behaviour emerges. On stretching the gel, the gel forming
micelle, (B) crystals, align parallel to the orienting influence as
expected from the affine stretching of a network. The platelet (A)
crystals however may orient differently. This distinction can be
conclusively followed as, in light of what has been said above, the
(A) and (B) crystals have recognizable X-ray finger prints. (Reflec-

tions associated with (B) are broader and at slightly larger angles; for further distinctions see chapter 4.) The "anomalous" orientation of (A) crystals consists of the fact that the molecular axes (c) become perpendicular to the stretching direction.*

We interpret such orientation effects as the result of the lamellae themselves aligning (i.e. as full morphological units) with the lamellar planes along the stretch direction (Figs. 3,4). The lamellae (A crystals) themselves need not be part of the network, but could be only loosely connected (i.e. non-load bearing) occlusions within it[6,7]. The fact that they can be even completely unconnected molecularly has been brought home by the analogous alignment of platelet-like zinc stearate crystals within PVC, present as stabilisers in the commercial material (Fig. 5).

There can be two variants to the above anomalous alignment of (A) crystals.
a) In addition to the lamellar plane a particular lateral crystal axis within this plane also aligns along the stretching direction. With PVC it is the a axis[6] (Figs. 2,3), with polypropylene it is the b axis[8] (Fig. 6). (Note, b axis orientation in polypropylene is unprecedented and seems to be a characteristic product of gel processing alone. The frequently reported anomalous orientation in past works on melt crystallised material is an a axis or a* orientation faintly visible in a heat treated gel sample, Fig. 6c, provided here for comparison.)
b) The plane of the lamellae align, as above, i.e. parallel to the draw direction, but without any preferential direction within the plane. In other words, the only orientation present is that of the lamellar normal which has become perpendicular to the stretch direction (Fig. 4). Examples are isotactic polystyrene[9] and polyethylene[10,11] (Fig. 7). (Polyethylene does gel but the corresponding gel belongs to Class II which is the consequence of stirring – see later.) According to a latest example PVC may also display effect b) when deliberately highly syndiotactic, hence highly crystallisable, fraction is incorporated into the crystallising solution (Fig. 8).**

---

*In most cases this was deduced from lateral, i.e. hk0, reflections in the X-ray diffraction patterns. However, even in these cases this assignment is fully supported by birefringence, e.g. negative birefringence in case of preponderant perpendicular (c) orientation, and by infra-red dichroic measurements[5,6].

**The distinction between the models in Figs. 3 and 4 is quite evident in the X-ray diffraction patterns (compare Figs. 2 and 8). It can be more rigorously demonstrated by pole figure representations not to be carried out here.

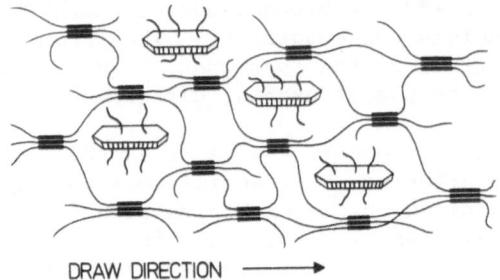

DRAW DIRECTION ⟶

Figure 3. Diagrammatic representation of a stretched gel containing
           lamellar crystals which are only loosely connected with,
           or totally unconnected to the network. The network form-
           ing micellar crystals align with their chain (c) axes
           along the draw direction, while the lamellar crystals
           align with their lamellar planes parallel to the draw
           direction. The latter implies that the chain direction
           within the lamellae is perpendicular to the stretch direc-
           tion. In addition, one of the lateral crystal axes, as-
           sumed here to correspond to the longest direction of the
           elongated crystal lamella, is also aligned along the draw
           direction. In the present works PVC and polypropylene cor-
           respond to this situation. In the former it is the a
           axis[6,7], in the latter the b axis[5,8] within the lamellar
           crystal which aligns along the draw direction.

     Thus the anomalous orientation effects represent a further mani-
festation of the distinctness of crystals A and B. The subclasses
a) and b) for A we tentatively associate with differences in the details of the lamellar morphology. Accordingly, highly elongated,
lath shaped crystals should fall under a) (the lath or ribbon directions aligning preferentially) while isodimensional crystals, like
the hexagonal i-PS, showing no preference laterally, fall in
category b).

3. Copolymer Crystallisation

     It is an observational fact of long standing that copolymers,
where at least one sequence is crystallisable, or where the crystal-
lisabilities of the comonomers are largely different, are particularly prone to set as gels when their solutions are cooled, under conditions where the corresponding homopolymers would precipitate as
disconnected platelets (e.g. [13,14,15]). Qualitatively it is obvious
why this should happen. Under conditions appropriate for the crystallisation of one of the comonomer species, this species would crystallise, nevertheless its average sequence length being too short for

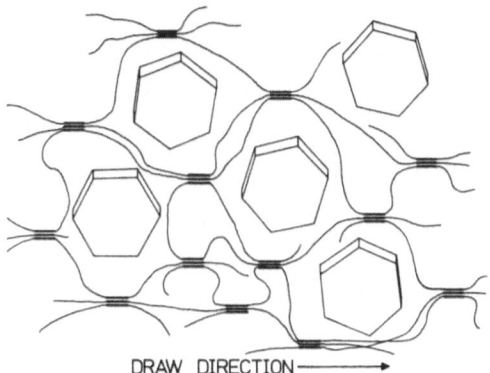

DRAW DIRECTION⟶

Figure 4.   Representation of a stretched gel containing lamellar
crystals as in Fig. 3.   The gel forming micellar crystals
are aligned as in Fig. 3, but the lamellar crystals, which
are here closely isodimensional platelets, align with
their lamellar planes only without preference of any other
crystal direction within this plane.   In the present works
isotactic polystyrene[9] (the hexagonal platelets were drawn
to represent this) and polyethylene[10] obey this
orientation scheme.

giving rise to chain folded platelets it has to associate with corresponding sequences from other chains, hence the crystal junctions,
micelles and ultimately networks which become manifest in the form
of gels.   Examples include various copolymers of polyethylene with
polyolefins which are of great current topicality[13,14].   Nature provides interesting examples, e.g. the seaweed polysaccharide alginate,
a copolymer of β-D-mannuronic acid and of α-L-guluronic acid[15].

In all these examples the copolymeric nature of the macromole-
cule was deliberately created by man or nature.   In the case of PVC
we have a copolymer of differing stereoisomeric entities as an acci-
dental consequence of the mode of synthesis.   (This will feature
further below.)   The above considerations are important from the
point of view of copolymer crystallisation per se, as it is the
dominant mode by which such copolymers crystallise.   The significance
of the process however, could be looked at in reverse: Is the gelation mode of crystallisation, when observed, a symptom of the copolymeric nature of the molecule?   This issue is particularly acute
when the gel remains stable, i.e. does not synerese.   If there are no
chemical interruptions along the chain why should the micelles not
keep on growing longitudinally until the whole gel collapses?   After
all, such a gel would not be an equilibrium gel, but is highly supercooled regarding the junction forming crystals.   Indeed, when we know

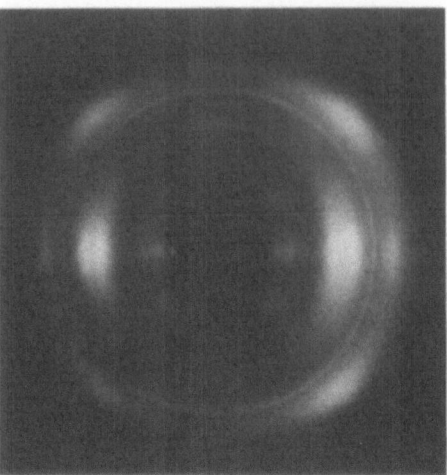

Figure 5.   X-ray diffraction pattern of a stretched and dried PVC
            (50°C polymerized) gel. Draw direction vertical. The
            sharpest reflections (at the lowest angles equatorial) are
            due to Zn stearate stabiliser crystals, demonstrating that
            even extraneous lamellar crystal matter will orient
            according to the schemes in Figs. 3 or 4[6].

a priori that the chain is chemically perfect, as in the case of
polyethylene (Class II, see below), the gels do synerese. In other
words, is the formation of a <u>stable</u> gel a sensitive indicator of the
copolymeric nature of the chain and/or of chemical (including stereo-
chemical) imperfections in general? To my knowledge there exists no
answer to this question.

4. Stereoregularity (PVC)

      The issue to be raised here follows directly from Chapter 3
above as applied to the example of PVC, a polymer of great interest
per se.

      It has been known for long that PVC gels from solution, and the
agency of crystal junction has been long suspected. However, in commercial material, the existence of crystallinity itself being under
dispute, e.g. [16], direct evidence for this was lacking. Our own
work provided this evidence by X-ray diffraction (e.g. Figs. 2,9).

      The most pronounced effect was the conspicuous nature of the
crystallinity even in commercial material as apparent by visual inspection of the X-ray diffraction patterns (Figs. 2,9). In fact

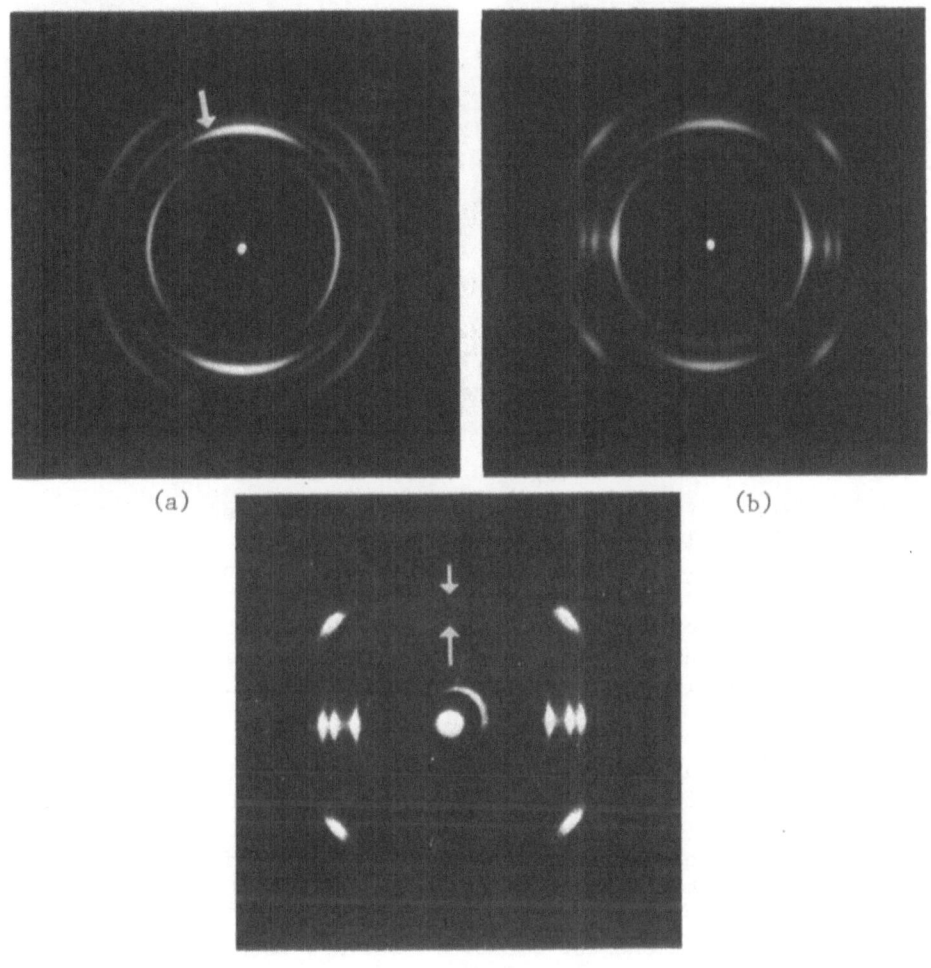

(a)

(b)

(c)

Figure 6.   X-ray diffraction patterns of stretched gels of polypro-
            pylene[5,8]. Draw direction vertical. a) Showing predomi-
            nantly b axis orientation (the strongest meridional 040
            reflection is arrowed). b) Showing a combination of b and
            c orientation. c) After heat treatment under load. The
            orientation is now nearly entirely of c axis type, but
            traces of b axis orientation (040 reflection, short arrow)
            and the here newly arising (but familiar from past works)
            a* axis orientation (110 reflection, long arrow) are
            nevertheless apparent.

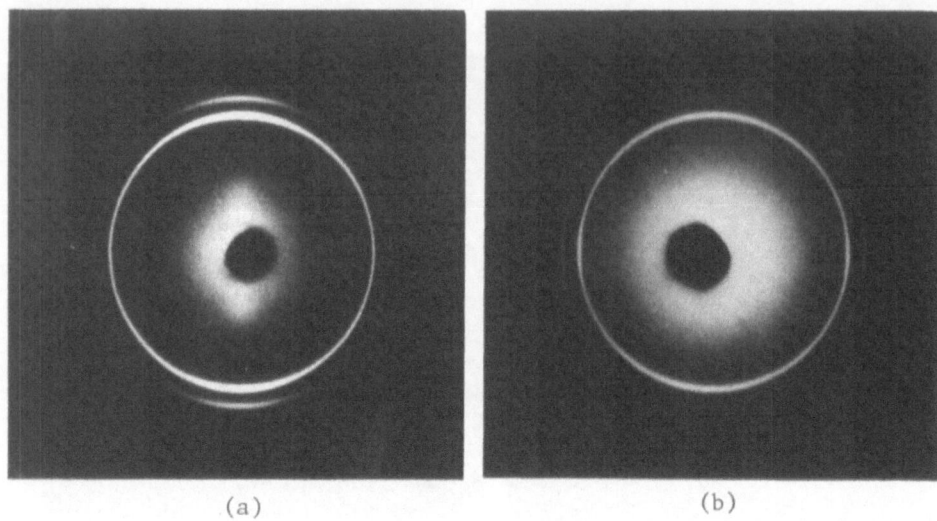

(a)   (b)

Figure 7. X-ray diffraction patterns of stretched and dried gels of
           polyethylene[10,11] (Class II - see Table). Draw direction
           <u>vertical</u>. a) Very slightly stretched, showing "perpen-
           dicular" fibre orientation interpreted as due to predom-
           inant amount of platelet material by scheme of Fig. 4.
           b) More highly stretched gel with a certain amount of
           "parallel" fibre orientation also present.

the study of these patterns greatly contributed to the establishment
of the above classification of A and B crystals.*

      PVC being largely atactic with some slight prominence of syndio-
tacticity, any crystallisation would correspond to that of a copolymer, the syndiotactic portion being the crystallisable component. So
far this follows from Chapter 3 above. The new issue, however,
arises as regards the quantity of syndiotacticity involved. By line
broadening measurements of the 002 reflection there would need to be
a minimum of 13 repeating units along the chain[17] even in the smaller
micellar (B) crystals - no such estimate has been made for the larger
A crystals to date. By NMR estimate, extrapolations from triad,

---

*In PVC the lamellar nature of the A crystals was inferred, rather
than directly observed by electron microscopy, at least so far. The
morphological identification of the A type crystals however was readily achieved on other polymers. Displaying identical orientation
anomalies, thermal features, etc., we ventured to attribute all
these effects to lamellar morphology in the morphologically so far
unexplored PVC.

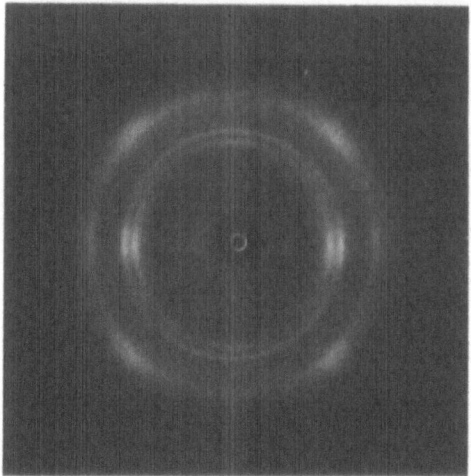

Figure 8.   X-ray diffraction pattern of stretched and dried gel of
            PVC (polymerized at -30°C) with a small amount of more
            highly syndiotactic low molecular weight material added[12].
            Draw direction vertical. We interpret it as corresponding
            to the equivalent of the scheme in Fig. 4 (as opposed to
            that in Fig. 3 attributed to the diffraction pattern in
            Fig. 2).

tetrad, etc., sequences on the basis of a Bernouillian statistic, the
proportion of syndiotactic sequences 13 repeat units long should be
negligibly small, about 0.07%. Bearing in mind that all these sequences have to "find each other" to form crystals, it will be clear
that a major inconsistency arises between the estimated tacticity
and the crystallographic evidence.

     There are two possible ways out of the dilemma: i) Bernouillian
statistics for tacticity distribution does not apply. Accordingly,
the higher sequences would be more frequent, the polymer being more
a block-copolymer of bunched syndiotactic units rather than a random
copolymer as hitherto assumed. ii) The isotactic sequences are
randomly distributed as assumed, but can be built into the
syndiotactic lattice.

     The implications of possibility i) are obviously far reaching
and are touching on issues such as the methodology of determination
of tactic sequences, and beyond it on the whole chemistry of PVC
polymerisation (see subclasses under 4 in Table 1). Preponderance
of syndiotacticity during the early stages of polymerisation, sub-
stantiated by experiments being performed in our laboratory[12,20],

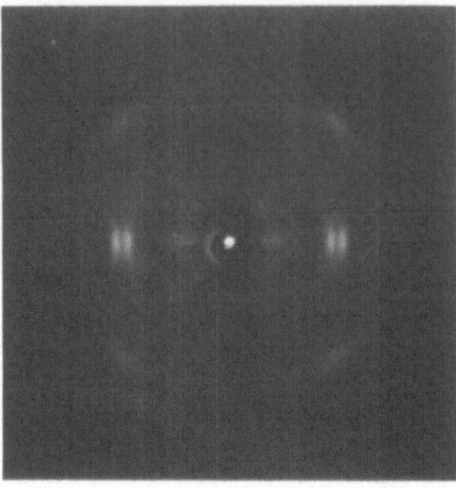

Figure 9.   X-ray diffraction pattern of a heat treated stretched gel
            of a commercial PVC, showing pronounced crystal reflec-
            tions refuting any notion of such a polymer being
            intrinsically amorphous[17]!

point to at least one obvious source of departure from non-
Bernouillian statistics of the syndiotactic sequence distribution
in the final product.

    Possibility ii) features in the literature prompted by much less
compelling effects than found by ourselves[19].   The model proposed
implies the feasibility of 60° rotations from the all transconfigur-
ation in the isotactic sequence which however, would violate the
principle of staggered bonds.*   While not dismissing this possibility
(in either event of i) or ii) something so far believed to be basic
would have to be abandoned or altered) this would clearly have far
reaching implications on the stereochemistry of PVC and polyolefins

---

*This drastic consequence is not immediately obvious from ref. [19]
which discusses the proposed "straight isotactic sequence" in terms
of traditional staggered, i.e. trans and gauche sequences. That the
actual model proposed requires ∿60° rotations, hence non-staggered
configurations, is however, apparent from the diagrammatic and pic-
torial representation of the model itself and from the basic contention that an essentially extended chain axial period of 5.0Å is maintained. The contention of an essentially TGT$\bar{\text{G}}$ (in conventional, not
the notation in ref. [19]) sequence being compatible with this, as
implied by the formal part of ref. [19], cannot be upheld, neither is
it consistent with the rest of the paper itself.

in general (item 6 in Table 1 with a different example on the same
general theme to follow in chapter 6).

As seen, the gelation phenomenon, apart from its interest per
se, has again led to enquiries far outside its original scope. In
this instance it serves as a sensitive indicator of the distribution
of syndiotacticity with all it entails. I think I have said enough
to "tick off" all items under 4 in our list (Table 1) as having been
raised and adequately documented and discussed for the present
purpose at least.

5. Plasticisation - A Technological Issue

The subject of plasticisation is of technological and practical
application and a forefront issue with PVC. It is while working on
PVC that it has come to our notice.

Presently we have formed the view that the essence of the technological process of plasticisation is to create a gel with, from the
mechanical point of view, load bearing junctions, and we affirm that
these junctions are micellar crystals in our category B. We confirmed that commercially used plasticisers are indeed gel forming agents
under our experimental conditions at the much higher solvent concentration of our experiments (99-90% as opposed to say 20% in technological plasticisation). While A crystals may or may not form under
technological conditions, it is the B crystals which carry the load
and make a plasticised object useful.

It is to be recalled that a purely amorphous polymer is not
expected to be plasticisable in the commercially useful sense, as the
lowering of T_g involved would turn it into a viscous liquid. To be a
structurally useful material, junctions are needed, and it has been
long suggested[21], even at the time when PVC was considered purely
atactic and amorphous, that such junctions in swollen PVC (as studied
at the time) may be tiny crystals. In the light of our present
experience on gel crystallisation, this now falls readily into place.

It is to be noted that the plasticisability of PVC comes about
through its ability to give rise to junction-forming crystals of
small size within an otherwise amorphous material. According to the
foregoing this should be possible by a contiguous, yet limited, number of sequences of stereoregular blocks within an otherwise atactic
polymer. If the blocks are too short, the material will remain essentially amorphous and will be either a glass or, above T_g, will
behave like a viscous liquid or creeping rubber (unless of course it
is chemically cross-linked). If the regular blocks are too long and
represent a sizable portion of the total molecule we get the usual
crystalline thermoplastic polymer. Thus, as far as the achievement
of a plasticised polymer is an aim in itself, the criterion by our

suggestion[18] would be to have just the right amount and distribution
of stereoregularity (syndiotacticity in the case of PVC), a criterion
which the current technological PVC happens to satisfy.

6. Chain Conformation of Polyolefins

This chapter concerns itself exclusively with isotactic polystyrene (i-PS). Historically, it was the crystallisation studies on
i-PS which directed our attention to the whole subject of thermoreversible gelation. Many of the previous points were in fact
established on i-PS, at least i-PS greatly added to the formation of
our present perspectives. Yet i-PS revealed a unique, additional
feature which may or may not be singular to i-PS (we do not know: it
may be quite general, it may be confined to polymers with aromatic
appendages, or to i-PS alone). Whatever the case, gelation crystallisation of i-PS has raised totally novel features and directed (or
diverted according to point of view) our enquiries to an unsuspected
further sphere.

The main point for the present chapter is that with i-PS the gel
forming crystals (B) and the lamellar crystals (A) are different, not
only as regards formation temperature, morphology, crystal size and
perfection (exact lattice parameters) as is the case with all the
other polymers referred to so far, but also as regards the actual
crystal structure in the traditional crystallographic sense, i.e. in
terms of molecular conformation and packing[2]. X-ray diffraction on
stretched gels reveals that it is the A crystals which possess the
traditional Natta-Corradini structure with the chains in the form of
a 3_1 helix while the structure of the gel forming B crystals is
totally different (Figs. 10,11,12). The most essential point as
regards this new structure of the B crystals is that the fibre
periodicity is close to that of fully extended all trans (TT) chains
(in fact, it is a slowly winding helix with twelve monomer units per
turn)[22]. By traditional view of vinyl polymers with substantial side
groups, extended or closely TT chain conformations should be steric-
ally impossible, hence the need for a trans-gauche (TG) sequence
which leads to the conformation of a 3_1 helix.

On recognizing a near extended TT conformation, the following
alternatives have arisen: i) Is the material fully isotactic, or is
it a copolymer containing a different isomer which is responsible for
the new crystal? ii) Can i-PS itself exist in a near extended
conformation, the classical texts to the contrary notwithstanding?

Our first suggestion[2] based on our belief in the inviolability
of the classical texts, favoured i), and in an attempt to match, both
the geometry of the diffraction patterns (layer-line spacings) and
the intensity of the reflections along the layer lines (note e.g. the
strong meridional at 5.1Å which requires the existence of a basic

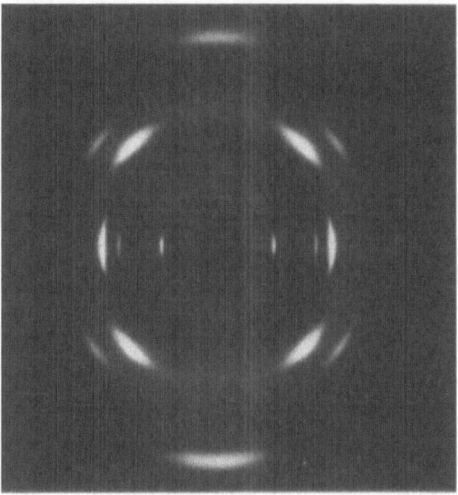

Figure 10. X-ray fibre diffraction pattern of isotactic polystyrene
(i-PS) with the traditional 3_1 helical (Natta) structure.
(This preparation was in fact obtained via the stretched
gel route followed by appropriate heat treatment - see
ref. [2].)

repeat of 2 monomer units without any halving, as e.g. a syndiotactic
unit would do), we suggested the existence of head-to-head, tail-to-
tail sequences (syncephalic) as the isomerically different comonomer
giving rise to crystals responsible for the new pattern. The match
achieved with the diffraction pattern was satisfactory, in fact the
most satisfactory to date[22]. Yet NMR evidence made this point of
view untenable, as according to the latter our polymer was 98-99%
isotactic. Accordingly, there is no latitude for the presence of any
other isomer in sufficient quantity to crystallise and give the new
pattern in observable strength.

While I personally do not regard this issue quite closed, we
turned to point ii) and reexamined the conformational possibilities
of pure i-PS. Rather confusingly, from the point of view of our
problems, conformational analyses from two different sources[23,24]
announced the existence of an energetically favourable near TT conformation, in addition to the traditional TG, the latter of course
giving the 3_1 helix for i-PS. At first sight this might have seemed
to provide the solution to our problem. In fact it did not, as the
particular small departure from the exact TT was in a sense as to
lead to closed rings (Fig. 14), a fact not directly apparent from
the analysis in ref. [23] which was on a dimer. Thus, the near TT

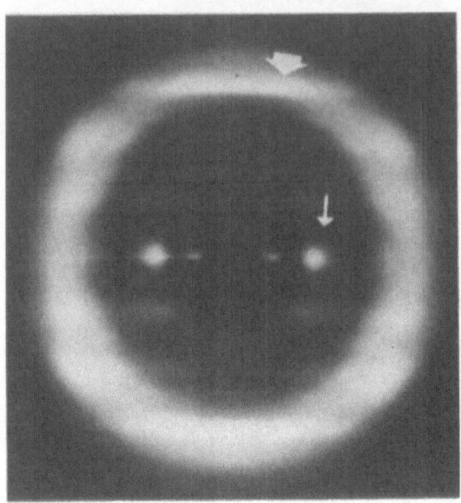

Figure 11.   X-ray diffraction patterns of stretched i-PS gel obtained
             from trans decalin displaying the novel crystal struc-
             ture. Note the weak odd layer lines at 30.6Å spacing
             (1st layer line arrowed) and the strong meridional at
             5.1Å (top arrow). Based on ref. [22].

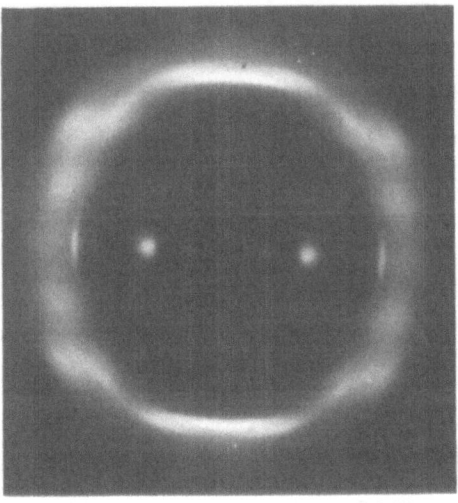

Figure 12.   X-ray diffraction pattern of stretched i-PS gel as in
             Fig. 11 with traces of the 3_1 helical structure also
             present (the sharp equatorial arcs)[22].

backbone in refs. [23],[24] could never lead to a fibre, not to speak of a crystal lattice. In fact, taking departures from eact TT in a different quadrant pair of the conformational map than was done in ref. [23],* the required extended near TT, as opposed to the ring shaped structure in Fig. 13 was achieved with the correct geometry to match the layer line spacing of the diffraction pattern (Fig. 15[25]). The energetic feasibility of this conformation was assessed by ourselves[25] and also by two other sources[26],[27] (the three parties in partial communication on this issue) and it was found that the structure in Fig. 15 with appropriate rotation of the phenyl groups was indeed possible and could, in fact, be the one with the minimum energy.

Nevertheless the problem is not yet fully solved. First, retrospectively, according to ref. [23] the quadrant pair which, as we now find as optimum, is of prohibitively high energy, and in fact was dismissed. One may ask why and how could such an even qualitative divergence of results occur. At present the answer seems to lie in the choice of non-bonded atomic radii in the calculation, where the exact value can be critical, a sobering reflection on the potential non-uniqueness of conformational analysis in general, even for such a comparatively simple molecule. More forward lookingly, however, even our extended near TT structure with minimum energy does not provide the full answer. It accounts for the geometry (layer line spacings) but not for the intensities. Some of the anomalies observed may be alleviated by appropriate side group rotation, but at least one feature seems to represent a fundamental conflict (weakness of 1st layer line in Fig. 11). Besides noting this, we ourselves have not done more about this intensity problem because we observed in the meantime that the intensities are not unique anyway but can depend on the solvent from which the gel has formed[28] (Fig. 13). It appears that there is no single, completely unique structure, even if there is likely to be a unique backbone geometry, the principal feature being the extended near TT chain backbone.

I shall not dwell on details of this issue further besides stating that the study of gelation has led to profoundly new issues relating to chain conformation of polyolefins, and with it to a reassessment of a subject which appeared already well nigh a closed chapter. Beyond it, the work has raised some issues on the power and validity of conformational analysis in general as briefly hinted on in the foregoing.

In addition, and beyond the problem of the true nature of the new polyolefin structure, is a more overriding question: why is the newly observed structure associated with gelation, and conversely,

---

*That is +23°, +11° as opposed to +23°, -11° in our notation; the sign convention in ref. [23] was different.

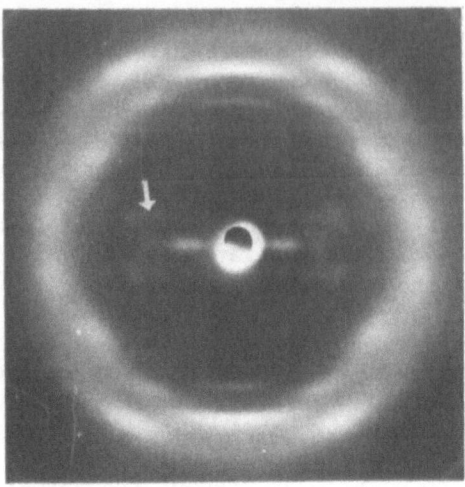

Figure 13. X-ray diffraction pattern of stretched i-PS gel obtained
            from cis decalin. The layer line periodicity is again
            30.6Å as in Fig. 11, but now the 1st layer line (arrowed)
            is comparatively stronger[28].

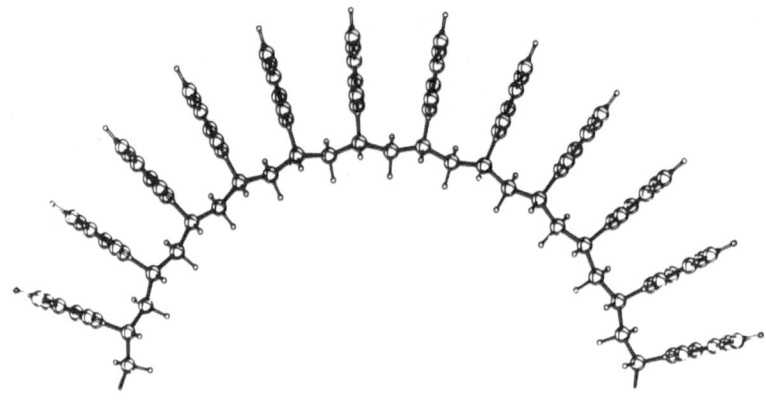

Figure 14. Projection of a section of i-PS in near all-transconform-
            ation with ψ_1 and ψ_2 (torsion angles around main chain
            C-C bonds) of opposite sign. This configuration, taken
            from ref. [24], traces out a large circle.

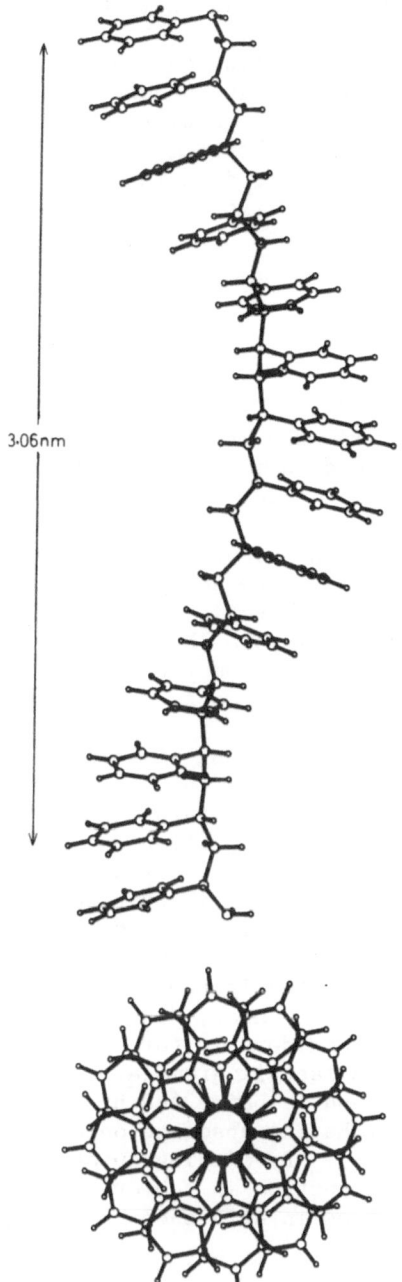

3·06nm

Figure 15. Projections of i-PS in near all-trans configuration with
ψ_1 and ψ_2 of equal signs leading to a highly extended
helix. This accounts for the geometry underlying diffraction patterns such as Fig. 11. Top: side view;
bottom: view along the helix axis.

why does gelatin-crystallisation, and only this mode of crystallisation, yield this structure? If I mention that there is evidence,
both old and new, of changes in solution conformation of polystyrene
in the temperature range where gelation occurs, affecting all solution properties, in some cases in a dramatic manner[2,29,30,31], I
could well add even the classical studies of polymer solutions to the
list in Table 1, as affected by, and as interacting with the subject
area under discussion. Nevertheless at this point I shall draw the
line.

CLASS II GELS

General Background

     Quite independently from the studies described above, gelation-
crystallisation has emerged also in a different area of activity with
which we happened to be associated. This was the study of flow
induced crystallisation, the origin and structure of the resulting
crystals (shish-kebabs) and the associated endeavors to attain high
modulus, high strength fibres. The activities of three separate
groups converged on the same subject: Smith and Lemstra (DSM),
Pennings (Groningen) and ourselves. The views of all parties concerned in this fluid subject do not match in every respect; here I
shall provide our own even if in utmost brevity.

     We have been engaged in producing high modulus polyethylene
fibres by the Pennings-Zwijnenburg method[32], which as will be known,
is carried out in a Couette type apparatus. Here, in our mode of
operation, the cylinder is set in rotation at a temperature which is
too high for crystallisation to occur, say at 135°C (see thermometer
scale in Fig. 16a). The temperature is then being lowered to within
the range marked "surface growth" in Fig. 16a. This is still too
high for crystallisation to set in spontaneously even on stirring.
Yet when a seed fibre is inserted in the solution, this when brought
in contact with the rotating cylinder, "catches on" so to speak, and
if pulled (it can support considerable force at the point of attachment) the fibre extends, in fact it grows at the tip of attachment.
This growth is continuous and is being fed by the molecules from the
solution, presumably after having absorbed onto the cylinder (entanglement adsorbtion layer). By winding up at the other end a continuous method of fibre production is achieved yielding some of the
strongest and stiffest fibres in existence (modulus up to 160 GPa,
tensile strength ∿ 4 GPa).

     In the course of experimenting by the above method we made the
following observation of significance for the present subject. To
facilitate conveying the essentials the general situation will be
described in terms of a specific example.

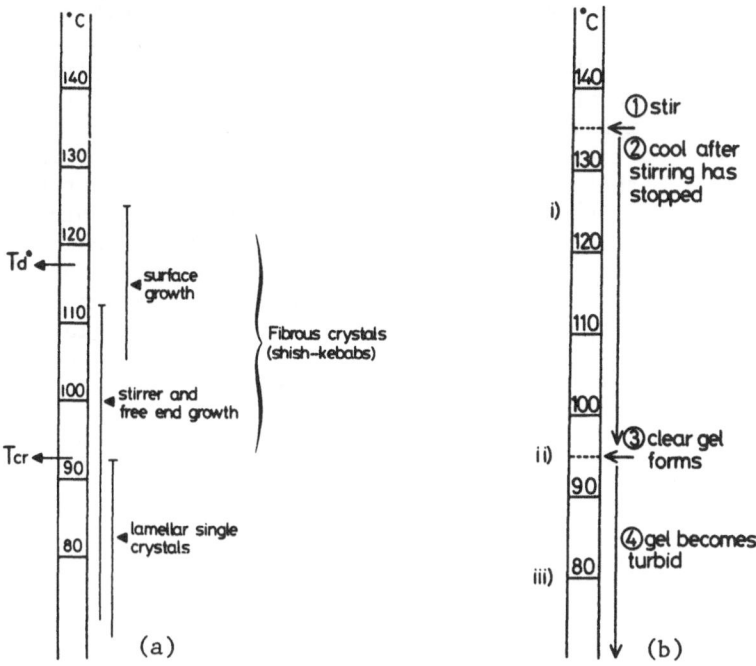

Figure 16. Temperature scales indicating conditions of: a) crystal-
            lisation, b) Class II gel formation of polyethylene in
            xylene and (in this respect very similar) decalin
            respectively. (For details of a) see ref. [34]; b) is
            based on refs. [33] and [44].)

    Look again at the thermometer in Fig. 16. At say 135°C abso-
lutely nothing happens. Neither are there any visible changes on
agitation, i.e. on rotation of the cylinder, at that temperature.
However, if agitation is brought to a halt and the solution thereafter (now in a quiescent state) is being cooled it sets as a transparent gel at a temperature of about 95°C (Fig. 16b). On further
cooling the gel becomes turbid owing to the usual precipitation of
lamellar crystals.

    At first we shall comment on the practical consequences of this
observation covering point 7 in the list of Table 1, and then proceed
to the more fundamental implications for oriented crystallisation and
for the gelation itself (points 8 and 9 in Table 1).

7. Gelation and Fibre Formation

    It has been found that only solutions such as gel in the manner
of the preceding chapter ('general background' under Class II) give

rise to fibres by the Pennings-Zwijnenburg surface growth method
(Fig. 17). The implication is that in the surface growth method
the fibre is actually being produced from a gel[33,34]. Stirring produces gel precursors which are probably gel particles of limited
extension at the high temperature in question (see later), and even
as far as these particles would cohere into a gel at sufficiently low
temperatures (see thermometer, Fig. 16), they would be kept broken up
by the continuing agitation. It is these gel particles which adhere
to the roller and which are stretched out by means of "hooking on" to
the seed fibre in the surface growth method. (Following this hypothesis, the adhesion of gel particles onto the roller was indeed directly identified, thus accounting for the "adsorbtion entanglement
layer" postulated originally to account for the surface growth
method[33]. Accordingly, the source of the production of ultra high

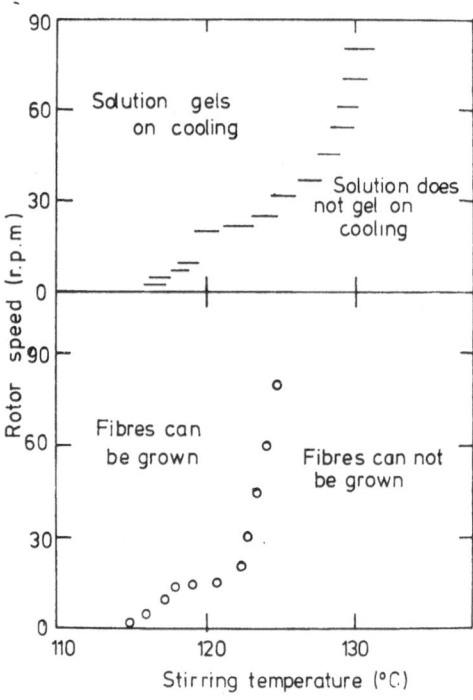

Figure 17.   An example of correlation between stirring induced gel-
             ling ability and the limits of growing fibres from the
             same solutions of polyethylene by the surface growth
             method[33].

modulus fibres is the stretching of gel. The roles of the adhesion of gel on to the roller, and that of the seed fibre is now seen to lie in the fact that they provide an effective means for gripping the gel.

The recognition that the route to high modulus fibres is the stretching of gels (or conversely, gel-stretching provides a way to high modulus fibres) leads to the design of more direct methods for producing such fibres: namely to produce gel intentionally and stretch it. One such explicit gel stretching method is that develop- ed into a continuous process by Smith and Lemstra[35] (first announce- ment in a paper by Pennings, Kalb, Smith and Lemstra[36]). It involves the passing of a solution through an orifice, cooling it subsequently, when a gel results due to the chain orientating effect of the elong- ational flow in the constriction (which in our view replaces the chain orienting effect of the agitation in the Couette). The result- ing gel can then be extended while still containing solvent or as completely dry, leading to fibres of ultra-high modulus.

A separate announcement of a new method for high modulus, high strength fibre production is via the porous fibre production method[37]. This consists of spinning a fibre from a hot paraffin solution during cooling, followed by extraction of the paraffin. The product is a highly "porous" fibre which can be drawn to very high draw ratios giving rise to ultra-high modulus and strength. Although not expli- citly stated in that work, in our interpretation such a system should also have gone through a gel state.

Thus, the common denominator of the different methods of high modulus fibre production involving solution processing is accordingly the gel state arising at some stage of the processing. Why the gel state in general should favour production of high modulus fibres is to be envisaged in a simplified sense as follows. It is well estab- lished in fibre science that, on preparing fibres by the usual drawing process, the modulus and strength increases with draw ratio. In traditional practice the achievable draw ratio is limited (by breakage) to what is conventionally, and misleadingly, termed "natural draw ratio", corresponding roughly to extensions of $\sim$ 6X. Here the chains are already fully oriented (but not yet fully stretched out).* It is a comparatively recent trend, that further improvements in moduli can be achieved by extending the range of draw ratios, presently up to 30-40X. This latter process is being termed "ultra drawing". It is in this way that ultra high moduli

---

*This distinction becomes meaningful if one considers the chain folded nature of crystals. Chain folded structures may align with the stems oriented in the draw direction while the chains remain largely folded. The result will be an "oriented" structure without much chain extension.

have been achieved in many cases with a unique relation between
modulus and draw ratio. It is envisaged that as a result of ultra
drawing, the chains are not merely oriented but in fact become fully
stretched out.

The method to chain extension just described is along the traditional route of cold drawing of the solidified thermoplastic material
(route 1). Another route (route 2) to high moduli is to stretch the
random chains first, best achieved in solution by flow methods, and
enable the fully stretched chain to crystallise subsequently so as
to form fibres. The surface growth method has been a development
along this route 2.

Now, the newly introduced or recognized gel stretching method.
seemed to bring us back to route 1, namely to the stretching of
material which has already crystallised. The advantage of the gel
state is its overall looseness (or porousness) which allows the
achievement of high extension ratios before break, the prerequisite
for high moduli. The chain alignment prior to crystallisation, and
the various agitation and flow methods to achieve it (which have
constituted route 2 originally) seem, from our new perspective, to
have the function to provide gel formation, and this by creating or
promoting the formation of junctions of a special kind (see below).
Thus in the light of these new developments the newly recognized gel
route to high modulus fibres embodies features of both routes 1 and 2
above. In technological terms it seems to be a combination of
spinning (of the "wet" variety) and drawing. In scientific terms it
is a combination of the stretching of random chains which then
induces crystallisation, and of the deformation of the already
solidified state coupled with strong elements of network stretching.

Some latest claims even dispense with the need of flow[38]. By
crystallising chains of sufficiently high molecular weight from
solutions of sufficiently high concentrations, the precipitate itself
acquires a cohering gel consistency which can then be drawn to high
draw ratios. This is in line with the above argument, in as far as
that once a stretchable gel is attained this will be advantageous for
fibre production. From the practical point of view the fundamentals
of how such a manipulable gel arises may not be of first importance,
nevertheless the basic science of why and how agitation promotes the
gel state remains a forefront issue to which we shall now turn.

8. Flow Induced Crystallisation

This issue clearly enters in the Class II gel formation. By
conventional conceptions, elongational flow stretches out chains
which then crystallise by forming·fibres with platelet overgrowth
(shish-kebabs). In most past studies chain stretching was achieved
by various flow methods creating elongational flow fields, e.g.
ref. [39]. However, it was known that such stretching can also be

achieved by the static methods of stretching lightly cross-linked
networks with the same result, namely leading to shish-kebab type
crystals. The latter has been long documented in single component
(molten) systems such as cross-linked polyethylene[40],[41] and isoprene
or butadiene based elastomers[42], and more recently, explicitly in
swollen networks of chemically cross-linked polyethylenes[43]. The new
developments, namely the recognition that agitation of solutions
leads to thermoreversible gels, modifies the picture only that much
that clearly many situations which previously were believed to
arise through stretching out of individual chains by hydrodynamic
means appear to correspond more to mechanical stretching of networks.
In other words, many of the shish-kebab preparations studied in the
past may have resulted along the network route. Clearly the new
recognitions now raise the question of when and how it can be ensured
that shish-kebabs form genuinely from flow induced prealigned isolated chains, an enquiry which is for further research to pursue.
In brief, no contradiction has arisen, merely the awareness of a
distinction where previously no dividing line was believed to exist.

9. Origin and Nature of Flow-Induced Gels

This is probably the most forward looking fundamental aspect
under Class II. Let us reconsider what is observed. There are 3
stages: i) Agitation without visible gel formation, which nevertheless induces the gel formation subsequently. This can occur as
high as 135°C on our temperature scale. ii) The clear gel forming
from quiescent solutions around 95°C as a result of i). iii) Turbid
gel on cooling to room temperature under quiescent conditions.

i) _The effect of the initial agitation._ As a first order
diagnosis of something having happened on agitation, say at 135°C, is
the occurrence of gelation on subsequent cooling under quiescent
conditions after agitation has stopped.*

Accordingly, the state of the solution must have been affected.
If not an actual gel, at least some junctions must have formed such
as facilitate the establishment of a network on cooling. What are
these junctions? Two possibilities might arise: a) entanglements,
and b) crystals. The most obvious way to distinguish between them
seems to be through their life time; are they temporary or permanent?
To answer this question the solutions were stored isothermally after
having been stirred at the appropriate pre-gel temperature (say
135°C). The findings were as follows:

---

*Control experiments are most important. Accordingly, a solution of
the same composition without preceding agitation should not gel or
crystallize otherwise at 95°C. When cooled below 90°C it should produce the usual turbid suspension (not gel) due to formation of
particulate platelet crystals[33].

a)   The junctions are of limited life-time. On sufficient
storage time at 135°C (say) the gel promoting effect is lost.
b)   The storage time needed to lose the memory of the agitation
was in terms of hours, more precisely depending on duration, time
and temperature of stirring.

We therefore have a most puzzling situation where the junctions
created by the agitation are far too long lived to be envisaged as
mere molecular entanglements, which should disappear within milli-
seconds in a comparatively dilute solution. However, they are not
permanent enough to be considered as usual stable crystals. Even if
the crystals were only stable in the flow field one would hardly
expect them to survive for hours on cessation of flow, particularly
as at that stage they must be minute (no visible turbidity whatso-
ever). Nevertheless, facing the two alternatives, some fibrous,
junction forming crystallisation, intrinsically unstable (in the
absence of flow) yet surprisingly long lived seem more probable.
Clearly what happens in Stage i) is the key to the whole subsequent
behaviour and hence must be basic. Further, it appears to be a new
phenomenon in its own right.

It will only be mentioned that a host of viscosity effects are
observed at this stage. Thus the viscosity goes through a maximum as
a function of stirring time[44], and gel formation (type ii) on subsequent cooling after stirring has stopped only takes place once this
peak in the viscosity v. time curve has been reached. These viscosity effects confirm that changes have indeed taken place in the pre-
gel range (i). (Fig. 9 in Pennings salient paper[39] is related to
the effect in question but no justice can be given here to it in
this brief discussion.) Beyond the viscosity issue the effects in
question extend the implications of the present topics to the area
of solution rheology and entanglement theory (items 10 and 11 in
Table 1).

ii)   The transparent gel. This gel forms under quiescent
conditions at temperatures too high for normal quiescent crystallisation. It is clearly induced by the memory of the preceding
agitation under i). It is thermoreversible, but dissolves only at
surprisingly high temperatures; it may persist up to 140°C (in
decalin as solvent). By suitable electron microscopic sampling it
consists of fibrous crystals with a very fine scale shish-kebab
overgrowth. There can be little doubt that it is a network of high
stability fibrous crystals and contains elements of the much studied
shish-kebabs which here act as junctions. Thus what is observed here
comes closest to the products of the previously studied flow induced
crystallisation but in a somewhat different and certainly novel
content.
iii)   The turbid gel. As the turbidity appears on cooling to
room temperature, the corresponding gel now contains all the polyethylene which is capable of precipitation. The principal

components responsible for turbidity are lamellar crystals. These
are of two types.
      a) Lamellae which precipitate epitaxially onto the gel forming
fibres, causing and/or enhancing the shish-kebab appearance of the
latter. By our classification of shish-kebabs we would identify them
as the source of macro shish-kebabs[34]. Thus, these lamellae will be
parts, but not specifically load bearing (or connective) elements of
the network.
      b) Lamellae which precipitate on their own. They may be totally
independent suspended platelets such as would form in the course of
the usual single crystal preparation, or may possibly have loose
molecular connections with the network. In neither case do they
significantly contribute to the network formation per se.

      It is to be noted that gels under iii), at least with poly-
ethylene, are prone to synerese. Nevertheless the gel does not
completely collapse. There will be a stable swollen gel phase in
apparent equilibrium with a liquid sol phase.

      Heating of gels iii. Here the whole process involved in the
formation is reversed. At about 96°C the turbidity disappears, the
gel clears and thus the gel under ii) is reestablished. Gel ii) then
melts at more elevated temperatures, 130°C and above, as already
quoted.

      Stretching of gels iii. On extending such turbid gels first, at
low extension ratios, anomalous orientation sets in, indicating that
a portion of the crystals, which in polyethylene can be overwhelming
majority, aligns with the chains perpendicular to the stretch
direction. This is identical to the effects described under 2 (in
connection with Class I) corresponding to the alignment of lamellae
which are unconnected with, or only loosely enmeshed in the network.
(Fig. 6, in fact, was taken from the present example.) At this stage
also the birefringence is negative[5]. On higher extension ratios the
whole system becomes "normal", i.e. c axis oriented. The ultra high
extension ratios with the associated ultra high moduli described
under 7 above are in practice achieved by stretching gels of the present type iii. The fact that the whole material, and not only the
network elements, convert into fully stretched chains, must mean that
the loosely embedded lamellar elements become also involved and
eventually fully extended in a way which is not clear at present.

AN IMPORTANT DISTINCTION BETWEEN GELS OF CLASS I AND II

      Class I gels form at very high supercoolings, higher than the
normal particulate lamellar crystals of the same polymer. In
contrast, Class II gels form at very low supercoolings, lower than
the formation temperature of the lamellar suspension. Similarly, as
might be expected, Class I gels melt at lower and Class II at higher

temperatures than the corresponding lamellar crystals. The latter
difference is particularly conspicuous when turbid gels of Class I
and Class II are heated. In the case of Class I the gel melts first
leaving a turbid suspension which then clears at some higher
temperature[2], while in case of Class II, first the turbidity clears
leaving a transparent gel, which then transforms to a liquid on
further raising of temperature[44].

It follows that while both Classes I and II are thermoreversible
and have many properties in common, the junction forming crystals are
different in the two. In Class I they are smaller and less perfect
than the usual lamellar crystals and are likely to be micellar
arising at very high supercoolings as described earlier. In
contrast, the opposite is true for Class II junctions which are
likely to be fibrillar, being the products of stretching induced
chain alignment, either directly or as a consequence of some still
unexplored intermediary state (state i above).

If crystallised from sufficiently long chains at sufficiently
high concentration it is possible to obtain gels which involve no
novel morphological crystal element, but consist of the usual lamellae with sufficient number of chains incorporated in two or more
lamellae to create an overall connectedness. (This may be associated
with the existence of chain overlap, hence intrinsic entanglement, in
the solution state before precipitation[38].) From the point of draw-
ability, such gels may well be equally usable for attaining ultra
high chain extensions by virtue of their overall looseness (see 7
above) nevertheless, besides possibly throwing some light on entanglements, coil overlap, etc. prevailing in the solution prior to
crystallisation, which is of course of interest in its own right
(see entry 11 in the Table), it has no qualitatively new message to
convey.

RELATION TO TRADITIONAL GEL STUDIES

Having reached this stage, we have "ticked off", so to speak,
all items in Table 1, I hope justifying the initial claim that the
study of thermoreversible gels embraces, or impinges upon, such an
astonishing variety of topics as listed in this table.

Nevertheless, I realize that experts on gels may well feel
dissatisfied. They may well ask, how does all this relate to the
traditional discipline of gel studies? It may well appear that a
connection, if there is such, is loose and qualitative. The reason
for this is two-fold.

The first reason is more historical and subjective. We (speak-
ing for myself and colleagues only) did not set out to investigate
gels as such, neither were we equipped, instrumentation and know-
ledgewise, to do so. We stumbled upon gels in the course of a large

variety of investigations with different objectives. In some cases
these objectives themselves appeared at first sight totally disconnected issues, with gelation emerging as a common thread subsequently. In other cases, the occurrence of gelation, often observed
accidentally, proved signposts to puzzling problems which were frequently outside the scope of our intended activities to begin with.
These we then often embraced as a result and thus found their way
into our table. Whichever the case, in no instance was the study
of gelation the objective as such.

The second reason is more objective and fundamental. The
traditional and well formalized studies on gels deal essentially
with equilibrium systems where the junctions are more or less
permanent. Here the gel state can be treated by equilibrium thermodynamics, involving the partitioning of phases, equilibration between
osmotic and elastic forces, etc. The gels of our present concern
are on a whole in a non-equilibrium state, where the formation of
junctions to a greater or lesser degree in a reversible manner is
itself a variable. In these cases the system is usually in a state
of physical change during examination.*

It is possible of course that within more narrowly circumscribed
conditions a state of quasi-thermodynamic equilibrium could be specified even for our gels in which case the traditional quantitative
treatment of swelling equilibrium may well be applicable, thus
linking the multitudinous findings listed here to the formalised
discipline of gels. These possibilities are being currently
explored.

ABSTRACT

The phenomenon of thermoreversible gelation has come to our
notice in a number of different contexts in the crystallisation of
polymers. The common feature is that the crystals themselves act as
junctions leading to a network and hence must be of micellar type.
The gels thus arising form two distinct classes. Class I: These
form under quiescent conditions at very high supercoolings in
polymers which are sufficiently super-coolable, below the formation
temperatures of the traditional chain folded crystals. Class II:
Here gelation is induced by preceding agitation (not necessarily
gelling during agitation) and it occurs at low supercoolings, in
fact at temperatures which are above the usual lamellar crystal
formation. Class I gels have a lower, class II gels higher thermal

---

*It is important to remember that our gels are in a supercooled
state with respect to the melting point of the crystals which form
the junctions, and the final equilibrium, at least in the case of a
chemically regular chain, should be the ideal crystal, i.e. the solvent free system.

stability (dissolution temperature) than the conventional crystals.
Both Class I and II can coexist with lamellar crystals leading to
unusual orientation effects on stretching. Prominent examples of
class I are isotactic polystyrene, where the gel is associated with a
novel unexpected chain conformation of major interest for polyolefins,
and PVC, where it has led to unusually pronounced cyrstallinity with
consequences as regards accepted notions of its tacticity with
further implications for its chemical synthesis and plasticisability.
Class II gels are prominent in polyethylene and polypropylene, where
in polyethylene they are the source of the stiffest and strongest
fibres as yet attained.

REFERENCES

1. G. Rehage, *Prog. Colloid Polymer Sci.*, 57:7 (1975).
2. M. Girolamo, A. Keller, K. Miyasaka, and N. Overbergh, *J. Polymer Sci.*, *Physics Ed.*, 15:211 (1977).
3. Faraday Discussion No. 68 (1979).
4. H. Berghmans, N. Overbergh, and F. Gavearts, *J. Polymer Sci.*, *Physics Ed.*, 17:1251 (1979).
5. P. Hawkins, C.G. Cannon, A.G. Coombes, and A. Keller, unpublished observations.
6. S.J. Guerrero, A. Keller, P.L. Soni, and P.H. Geil, *J. Polymer Sci.*, *Physics Ed.*, 18:1533 (1980).
7. S.J. Guerrero, A. Keller, P.L. Soni, and P.H. Geil, presented at International Conference on PVC, Cleveland (1980), to appear in *J. Macromol. Sci.* B 20:161 (1981).
8. P.J. Lemstra, O. Acs, and A. Keller, unpublished works.
9. J.S. Shapiro and A. Keller, to be published.
10. A.G. Coombes, Ph.D. Thesis, Bristol (1979).
11. C.G. Cannon, unpublished communication.
12. S. Fulton and A. Keller, to be published.
13. R. Benson, J. Maxfield, D.E. Axelson, and Mandelkern, L., *J. Polymer Sci.*, *Physics Ed.*, 19:1583 (1978).
14. A. Takahashi, *Polymer J.*, 4:379 (1973).
15. E.D.T. Atkins, I.A. Nieduszynski, W. Mackie, K.D. Parker, and E.E. Smolko, *Biopolymers*, 12:1865 (1973).
16. R.S. Straff and D.R. Uhlmann, *J. Polymer Sci.*, *Physics Ed.*, 14:35 (1976).
17. P.J. Lemstra, A. Keller, and M. Cudby, *J. Polymer Sci.*, *Physics Ed.*, 16:1507 (1978).
18. S.J. Guerrero and A. Keller, paper presented at International Conference on PVC, Cleveland (1980), to appear in *J. Macromol. Sci.* B 20:167 (1981).
19. J.A. Juijn, J.H. Gisolf, and W.A. de Jong, *Kolloid Z.u.Z. Polymere*, 251:456 (1973).
20. S.J. Guerrero, Ph.D. Thesis, Bristol (1980).
21. T. Alfrey, N. Wiederhorn, R. Stein, and A. Tobolsky, *Ind. and Eng. Chem.*, 41:701 (1949).

22. E.D.T. Atkins, D.H. Isaac, A. Keller, and K. Miyasaka, J. Polymer Sci., Physics Ed., 15:211 (1977).
23. D.Y. Yoon, P.R. Sundararajan, and P. Flory, Macromolecules, 8:776 (1975).
24. L. Beck and P.C. Hägele, Colloid Polymer Sci., 254:288 (1976).
25. E.D.T. Atkins, D.H. Isaac, and A. Keller, J. Polymer Sci., Physics Ed., 18:71 (1980).
26. P.R. Sundararajan, Macromolecules, 12:575 (1979).
27. P. Corradini, G. Guerra, V. Petracone, and B. Pirozzi, Europ. Polymer J., 16:1089 (1980).
28. E.D.T. Atkins, A. Keller, P.J. Lemstra, and J.S. Shapiro, Polymer, 22:1161 (1981).
29. C. Reiss and H. Benoit, J. Polymer Sci., C16:3079 (1968).
30. A.M. North, A. Richard, H. Petrick, and Poh Beng Teik, Polymer, 21:769 (1980).
31. M.J. Miles, K. Tanaka, and A. Keller, to be published.
32. A. Zwijnenburg and A.J. Pennings, Colloid and Polymer Sci., 259:868 (1978).
33. P.J. Barham, M.J. Hill, and A. Keller, Colloid and Polymer Sci., 258:899 (1980).
34. A. Keller and P.J. Barham, Plastics and Rubber International, 6:19 (1981).
35. P. Smith and P.J. Lemstra, J. Material Sci., 15:505 (1980).
36. P. Smith, P.J. Lemstra, B. Kalb, and A.J. Pennings, Polymer Bull., 1:733 (1979).
37. B. Kalb and A.J. Pennings, Polymer, 21:3 (1980).
38. P. Smith and P.J. Lemstra, and H.C. Booig, J. Polymer Sci., Physics Ed., 19:877 (1981).
39. A.J. Pennings, Polymer Symposia, 59:55 (1977).
40. A. Keller and M.J. Machin, J. Macromol. Sci., B1:41 (1967).
41. A. Keller and M.R. Mackley, Pure and Appl. Chem., 39:195 (1974).
42. E.H. Andrews, Proc. Roy. Soc., A277:562 (1964).
43. A. Posthuma de Boer and A.J. Pennings, Faraday Discussion, 68:345 (1979).
44. K.A. Narh, P.J. Barham, and A. Keller, Macromolecules, 15:464 (1982).
45. S. Wellinghoff, J.Shaw, and E. Baer, Macromolecules, 12:932 (1979).

CONFORMATIONAL CHARACTERIZATION OF THE

SINGLE CRYSTAL SURFACE

Y.A. Chang and A. Hiltner

Department of Macromolecular Science
Case Institute of Technology
Case Western Reserve University
Cleveland, Ohio 44106

ABSTRACT

The surfaces of polyethylene single crystals have been selectively brominated, and the conformational sensitivity of the C-Br
infrared stretching modes has been utilized to characterize the fold
structure. From the analysis of low molecular weight compounds,
band assignments have been made for each of the six possible conformations of the two C-C bonds adjacent to the C-Br bond. The
gauche conformations predominate in the spectra of single crystals.
The calculated conformational populations were observed to change
with temperature particularly in the region of the dynamic mechanical
α relaxation. As a result of similarities observed in the infrared
spectra of brominated crystals and one of the model compounds,
bromocyclooctane, models for the chain fold have been proposed.
These consist of nine carbon atoms arranged in the crown or twisted-
crown structures. Both models conform well to the lattice
constraints of the {110} fold.

INTRODUCTION

The nature of the chain fold has been a problem challenging
polymer scientists ever since the first reports (1-3) of lamellar
single crystals of linear polyethylene grown from dilute solution
led to the hypothesis that the long-chain molecules had to fold in
order to be accommodated in the crystal. Several investigators
have attempted to use selective, nondegradative chemical modifica-
tion of the crystal surface as a probe of the fold structure.
Harrison and Baer (4) have shown that bromination of single crystals

results in selective substitution at the surface without modifica-
tion of the crystalline core. The infrared spectra of brominated
single crystals reported by Eguiluz et al. (5,6) differ significantly
from those of polyethylene which had been substituted randomly by
bromination in solution. It is well-known that the C–X stretching
modes are sensitive to the conformation of the adjacent C–C bonds.
Unfortunately, detailed interpretation of the brominated single
crystal spectra was hindered by the lack of reliable band assignments. The promising results of Eguiluz et al. (5,6) have been
extended in the present study. The infrared spectra of several
classes of model compounds are analyzed in order to make reliable
conformational assignments for the C–Br stretching bands. Subsequently the model compound studies have been used to analyze the
infrared spectra of brominated single crystals grown from a narrow
molecular weight fraction of polyethylene.

EXPERIMENTAL

    Linear polyethylene, Standard Reference Material 1484, was
obtained from the National Bureau of Standards. The weight and
number average molecular weights are 119,600 and 100,500 respectively
and Mw/Mn = 1.19. This polymer contains some vinyl end groups which
were hydrogenated by the method of hydroboration followed by pro-
tonalysis (7). The extent of reaction was determined from the
intensity of the infrared band at 909 cm^{-1} to be about 98% complete.

    Single crystals were grown isothermally in xylene at two temp-
eratures, 80°C and 70°C, by the self-seeding method (8). The suspension was filtered without drying and washed repeatedly with
bromobenzene until the suspension was free of xylene.

    The single crystal suspensions in bromobenzene were brominated
at 50°C by addition of elemental bromine in the presence of UV radiation as described previously (5). The brominated crystals were
washed with bromobenzene, acetone and anhydrous ether then dried
under vacuum at 40°C. The bromine content, measured by elemental
analysis, was 4.5% by weight for the crystal preparation grown at
80°C and 3.6% for that grown at 70°C.

    The single crystals suspended in bromobenzene were examined
morphologically before and after bromination. Specimens were placed
on carbon-coated grids, dried, shadowed with platinum-carbon at 30°
and examined with a JEOL 100B electron microscope.

    Specimens for infrared analysis were prepared by pressing a
thin mat of single crystals in an evacuated die at ambient temperature. Spectra were obtained with a Digilab FTS–14 Fourier-transform
infrared spectrometer using 200 scans of both the sample and reference. Spectra were obtained at temperatures up to 155°C by mounting

the specimen pellet between KBr discs in a Perkin-Elmer heating
apparatus. The temperature was controlled to within 1°C.

Bromocyclopentane, bromocyclohexane and bromocycloheptane were
purchased from Aldrich Chemical Co. and used without further purification. Bromocyclooctane and bromocyclododecane were synthesized
by treating cycloctene and cyclododecene respectively with hydrogen
bromide. Meso-2,3-dibromobutane and dl-2,3-dibromobutane were
prepared by bromination of trans-2-butene and cis-2-butene
respectively.

RESULTS AND DISCUSSION

A. <u>Single Crystals</u>

Typical electron micrographs of the single crystal preparations
are shown in Figure 1. The crystals grown at 70°C have a dendritic
structure with some spiral overgrowths, the thickness of the crystals is about 100Å. Crystals grown at 80°C have a truncated diamond
shape about 140Å thick and 8.5μ in length. Brominated single
crystals showed no morphological changes.

Infrared spectra of the brominated crystals in the 500-700 cm^{-1}
C-Br stretching region are shown in Figures 2 and 3. In both cases,

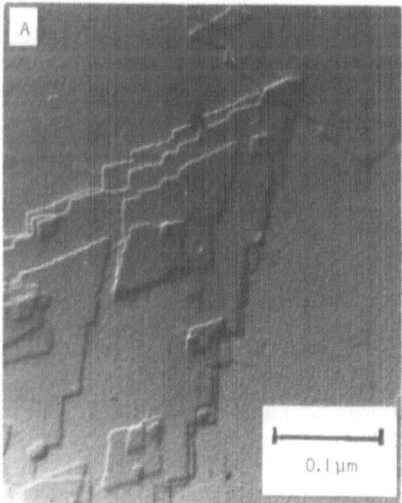
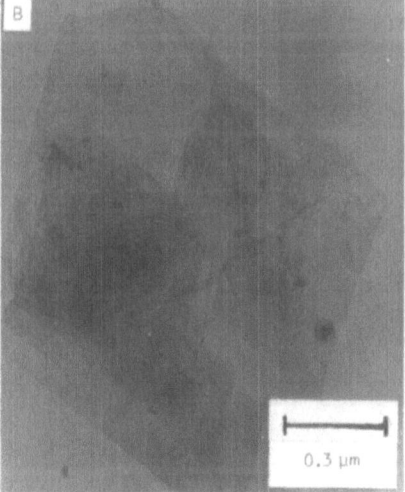

Figure 1.   Electron micrographs of (a) a polyethylene single crystal
grown at 70°C, and (b) a polyethylene single crystal
grown at 80°C.

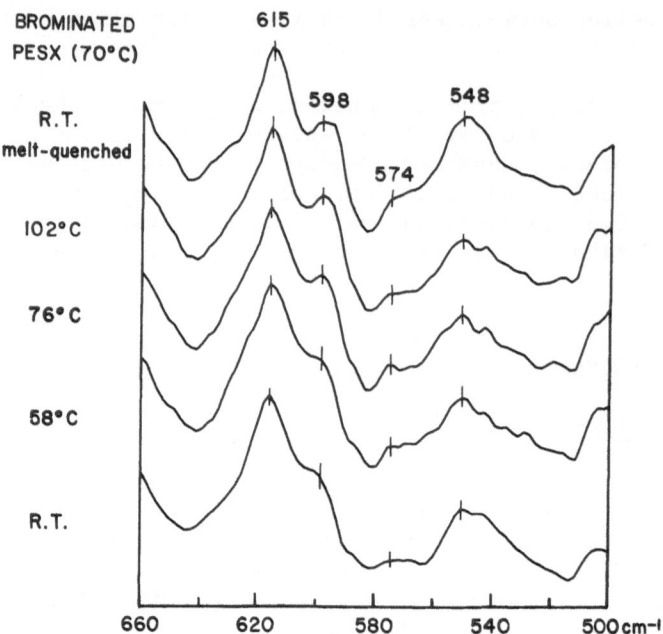

Figure 2.   FT-IR spectra of brominated polyethylene single crystals
           grown at 70°C. The spectra were taken at (a) room temp-
           erature, (b) 58°C, (c) 76°C, (d) 102°C, and (e) room
           temperature after the specimen was melted at 155°C.

a band at 617 cm^{-1} (615 cm^{-1}) is the most intense. Weaker bands
appear at 594 cm^{-1} (598 cm^{-1}), 574 cm^{-1}, and 548 cm^{-1}. Other bands
are observed at 674 cm^{-1}, 660 cm^{-1} and 630 cm^{-1} as shoulders on the
very intense crystalline band at 720 cm^{-1}.

Figures 2 and 3 also show the effect of temperature on the
infrared spectra. The intensity of the 574 cm^{-1} does not change
significantly with temperature and was chosen as an internal
standard. Relative to the 574 cm^{-1} band, the 617 cm^{-1} band decreases in intensity and the 548 cm^{-1} and 594 cm^{-1} bands increase
in intensity. The temperature effects are retained when the specimens are quenched from the melt. The absorbance ratios of the 548
cm^{-1} and 594 cm^{-1} bands relative to the 617 cm^{-1} band are plotted
in Figures 4 and 5. The largest changes with temperature are
observed in the 594 cm^{-1} band. In all cases, the intensity changes
occur in the 30°-80°C temperature range. No changes are observed
above 80°C.

Annealing experiments were performed at various temperatures
between 25°C and 123°C for 10 hrs. The spectra subsequently recorded at room temperature are essentially the same as the spectra

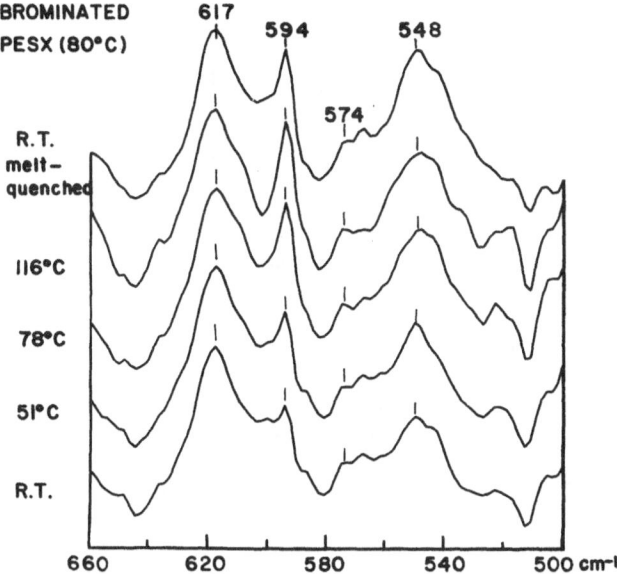

Figure 3. FT-IR spectra of brominated polyethylene single crystals grown at 80°C. The spectra were taken at (a) room temperature, (b) 51°C, (c) 78°C, (d) 116°C, and (e) room temperature after the specimen was melted at 155°C.

in Figures 2 and 3 obtained at the annealing temperature. However, upon prolonged aging at room temperature, the peak intensities change gradually (Figure 6) until the spectrum closely resembles that of the initial preparation. This is particularly evident in the gradual decrease in intensity of the 594 cm^{-1} band. The change is completely reversible and upon reheating the intensity again increases.

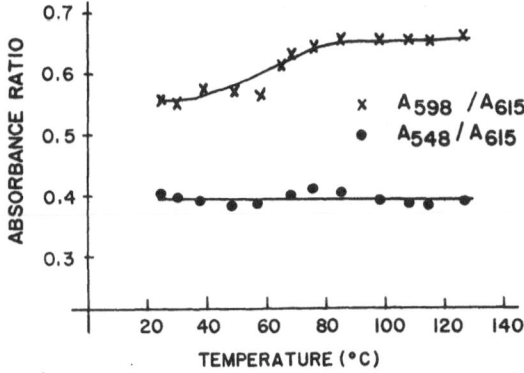

Figure 4. The absorbance ratio plotted as a function of temperature for brominated single crystals grown at 70°C.

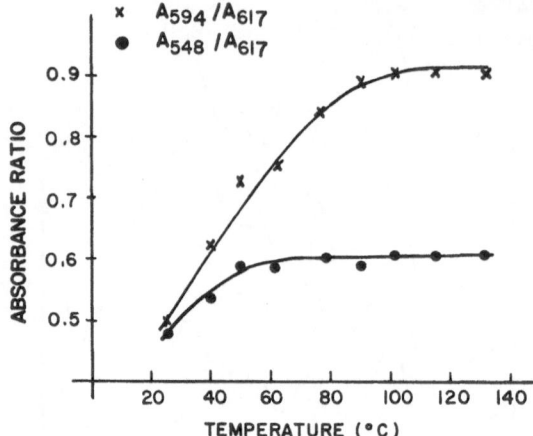

Figure 5.   The absorbance ratio plotted as a function of tempera-
            ture for brominated single crystals grown at 80°C.

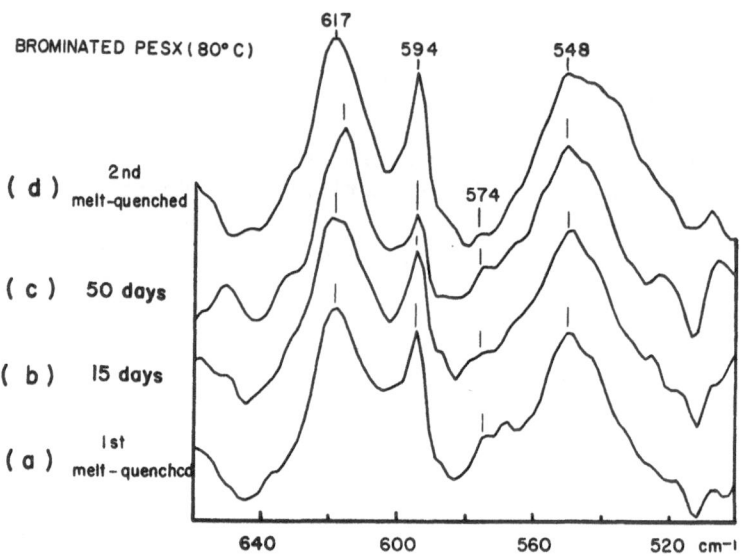

Figure 6.   FT-IR spectra of brominated single crystals grown at 80°C.
            The specimen was (a) melted at 155°C and cooled to room
            temperature, (b) aged at room temperature 15 days, (c)
            aged at room temperature 50 days, and (d) remelted at
            155°C and cooled to room temperature.

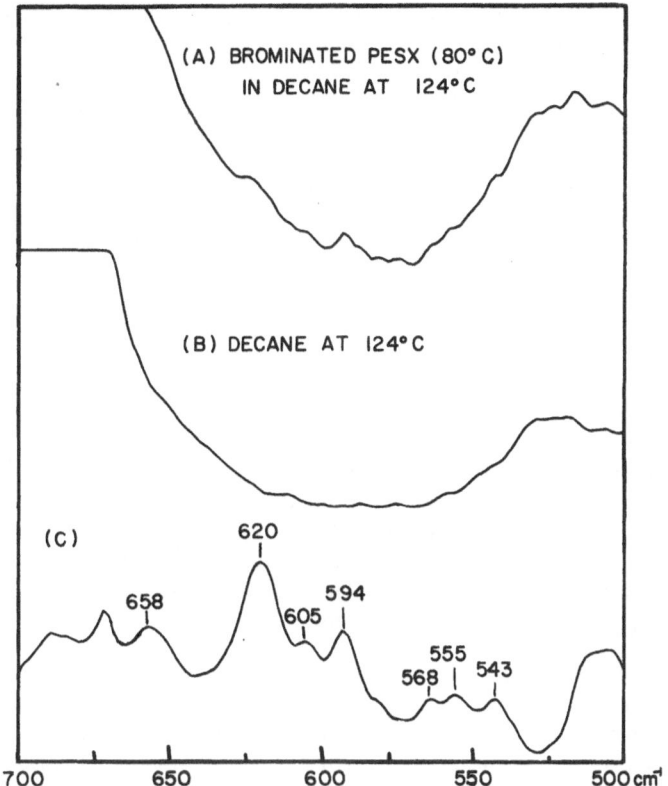

Figure 7. FT-IR spectra of (a) brominated single crystals grown
at 80°C dissolved in decane, the spectrum was taken at
the dissolution temperature, 124°C, (b) decane at 124°C,
and (c) difference spectrum.

The spectrum of the brominated single crystals dissolved in n-
decane was obtained by spectral subtraction of the solvent contribution (Figure 7). The difference spectrum shows the same bands as
the solid state spectrum although there are slight shifts in the
relative intensities.

B. Model Compounds

Band assignments for the single crystal spectra were based on
the analysis of spectra of three families of model compounds:
linear mono-bromoalkanes, vicinal dibromides, and mono-bromocycloal-
kanes. A detailed analysis of the spectra is presented elsewhere
(9).

Band assignments for the secondary C-Br stretching modes are
identified by the symbol S with subscripts which indicate the sub-

stituents in the positions trans to the bromine atoms (C,H,H'). The corresponding conformations of the two C-C bands adjacent to the C-Br bond are given in parentheses (T,G,G').

Three of the possible six C-Br stretching modes have been assigned from the study of the linear bromoalkanes (9). They are the S_{HH} (TT) mode at about 548 cm^{-1}, the S_{CH} (TG') mode at about 620 cm^{-1}, and the $S_{HH'}$ (TG) mode at about 574 cm^{-1}. The remaining higher energy conformations are observed in the cyclic compounds. They are of particular interest since the constraints which exist in ring compounds may produce conformations similar to those of a tight chain fold.

The spectra of the bromocycloalkanes are shown in Figure 8. The spectrum of bromocyclohexane shows sharp bands at 690 and 660 cm^{-1} and a weaker band at 505 cm^{-1}. In the liquid, the equatorial structure predominates, and the bands are assigned to equatorial and axial C-Br stretching modes respectively with the ring in the preferred chain conformation (10). Although the conformation of the C-C bonds is G'G in both cases, the substituents in the positions trans to the bromine atom differ, so the 690 cm^{-1} band is identified as S_{CC} and the 660 cm^{-1} band as $S_{H'H'}$.

The spectrum of bromocycloheptane is very similar to that of bromocyclohexane. The energetically favored conformation is the twisted-chair form with the substituent in an equatorial position. The bands are given assignments analogous to those of bromocyclohexane.

Cyclooctane is the smallest ring which can accommodate 4 carbon atoms in the trans conformation (11). Assuming the ring to take the crown form, the intense band at 619 cm^{-1} in bromocyclooctane is assigned to the S_{CH} (TG') conformation. Possibly the band at 590 cm^{-1} is associated with the $S_{HH'}$ (TG) conformation.

The spectrum of bromocyclododecane (Figure 9) has only two bands, a strong one at 618 cm^{-1} and a weaker one at 550 cm^{-1}. These are assigned to the S_{CH} (TG') and S_{HH} (TT) conformation respectively. From the relative intensities, it is apparent that the ring size permits isolated trans conformations but is too constrained for adjacent trans-trans conformations to be favored.

The C-Br stretching modes of compounds with more than one Br atom depend upon the relative positions of the Br atoms. When separated by one or more carbon atoms, the vibrational modes are not highly coupled and the spectra show the features characteristic of each C-Br bond (9). On the contrary, the stretching modes of adjacent or vicinal dibromides are highly coupled and often only a single infrared-active mode is observed for each conformation.

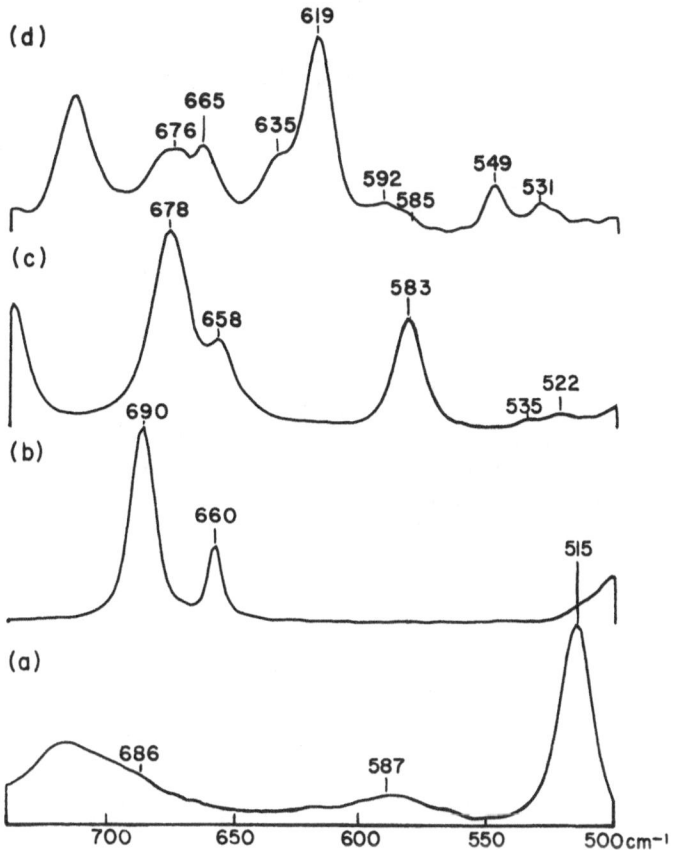

Figure 8.   FT-IR spectra of (a) bromocyclopentane, (b) bromocyclohexane, (c) bromocycloheptane, and (d) bromocyclooctane.

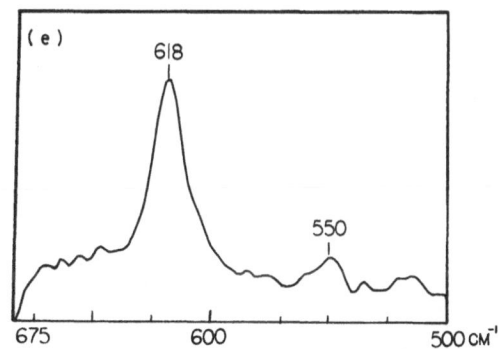

Figure 9.   FT-IR spectrum of bromocyclododecane.

Vicinal dibromides which are not terminal have two isomeric forms,
meso and dl. With the carbon atoms in an extended trans conformation, the meso form has the two Br atoms on opposite sides of the
molecule and trans to each other while the dl form has both on the
same side in one of two gauche arrangements.

The infrared spectra of the two forms of 2,3-dibromobutane are
shown in Figures 10 and 11. At ambient temperature, the meso form
has three bands at 638 cm^{-1}, 600 cm^{-1} and 552 cm^{-1}. Only the 552
cm^{-1} band is observed at -150°C in the crystalline sample. This
band is assigned to the extended trans conformation and the bands at
600 cm^{-1} and 552 cm^{-1} to the two gauche conformations (12). The
spectrum of the dl form at ambient temperature has four bands,
only the 650 cm^{-1} band disappears at -150°C. Unlike the meso form,
the trans conformation of the carbon atoms is not the most stable.
Conformational assignment of the bands cannot be made with certainty,
the 650 cm^{-1}, 560 cm^{-1} and 542 cm^{-1} bands are tentatively identified with the trans and the two gauche conformations respectively
(12). The 635 cm^{-1} band may be a Raman active symmetric stretching
mode.

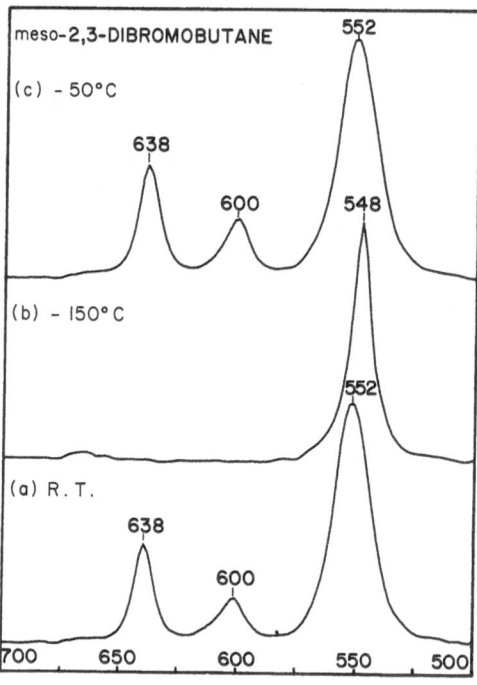

Figure 10. FT-IR spectra of meso-2,3-dibromobutane at (a) room
temperature, (b) -150°C, and (c) 0°C.

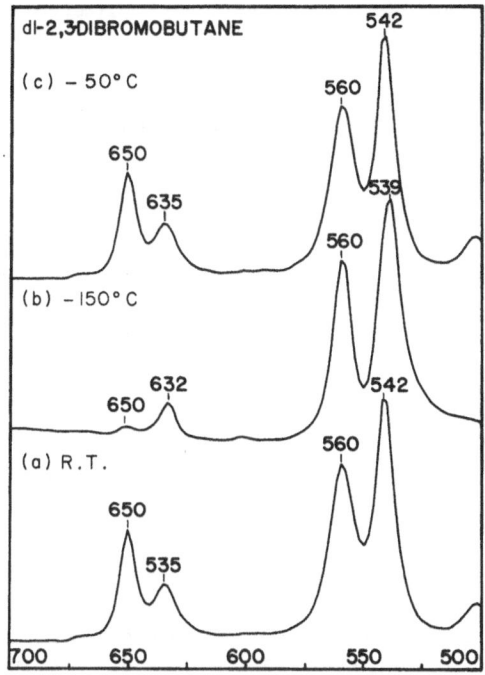

Figure 11.   FT-IR spectra of dl-2,3-dibromobutane at (a) room
temperature, (b) -150°C, and (c) 0°C.

C.   Chain Fold Models

     Conformational assignments for the bands in the spectra of
single crystals have been made on the basis of the spectral analysis
of the model compounds. All six of the possible conformations are
observed (Table I). The fraction of each conformation has been
calculated from the peak intensities. Although slight differences
are observed in the conformational distribution of the two single
crystal preparations, approximately half the brominated segments
are in the TG' conformation and 20% in the TT conformation. The
large percentage of gauche conformations clearly results from the
constraints imposed on the fold segments by the position of the
crystalline stems.

     The infrared spectra of brominated crystals closely resemble
that of bromocyclooctane in terms of the frequency and intensity
of the observed C-Br stretching bands. This suggests that the conformation of the chain fold may be similar to that of cyclooctane.
The stable forms of cyclooctane, crown and twisted-crown, are shown
connected with two zigzag crystalline stems in Figure 12. The
orthorhombic crystalline form (Figure 12(a)) is used, and folding

Table I. The Conformational Populations of Brominated Chain
Segments in Polyethylene Single Crystals.

Frequency (cm^{-1})	Assignment	Fraction (%, 80°C)[a]	Fraction (%, 70°C)[b]
674	S_{CC}(GG')	11	18
660	$S_{H'H'}$(G'G)	1	1
630	$S_{CH'}$(GG)	4	3
617	$S_{CH'}$(TG')	53	53
574	$S_{HH'}$(TG)	9	5
548	S_{HH}(TT)	22	20

[a] Single crystals grown at 80°C.
[b] Single crystals grown at 70°C.

is assumed to occur in the lattice plane of closest packing, the
{110} plane. The fold structure labeled A1 and A2 in Figure 12(b)
is the twisted-crown structure while that labeled B1 and B2 is the
crown structure. In both cases the fold contains nine carbon atoms.
In A1 and B1, the fold is viewed perpendicular to the {110} plane,
and parallel to the {110} plane in A2 and B2. The dihedral angles
measured from the molecular structure model give a fold conformation
for model A of -T'T*'G'TG*'GT'T*'-. For the type B the fold conformation is -T'T'G'T*'T*'GT'T-, where T = 165°, T' = 165°, G = 55°,
T* = 135°, T*' = 135°, and G*' = -85°. The distance between carbons 1 and 9 is 4.51Å for type A and 4.35Å for type B which is very
close to the experimental distance, 4.45Å (13). The {110} fold
model proposed by Corradini et al. (14) consists of nine carbon
atoms with -TGTGG*G'G'T- conformation, where G* = 91.6° is significantly displaced from the minimum energy state. This is consistent with the strained gauche (85°) conformation of the twisted-
crown model.

The carbon atoms accessible to bromination are carbons 4,5
and 6. Bromination of these sites would produce secondary C-Br
bands in TG', TG, TT, and GG conformations. The C-Br stretching
band assigned to each of these conformations is observed in the
spectra of brominated single crystals. The high intensity of the
617 cm^{-1} band (TG') suggests that this conformation is preferentially brominated, possibly due to steric factors. In addition, contributions to the spectra from brominated loose loops and cilla
are expected. Larger rings, such as cyclododecane, model the
loose loops. In this case, the bands due to two conformations,
TG' and TT, are observed.

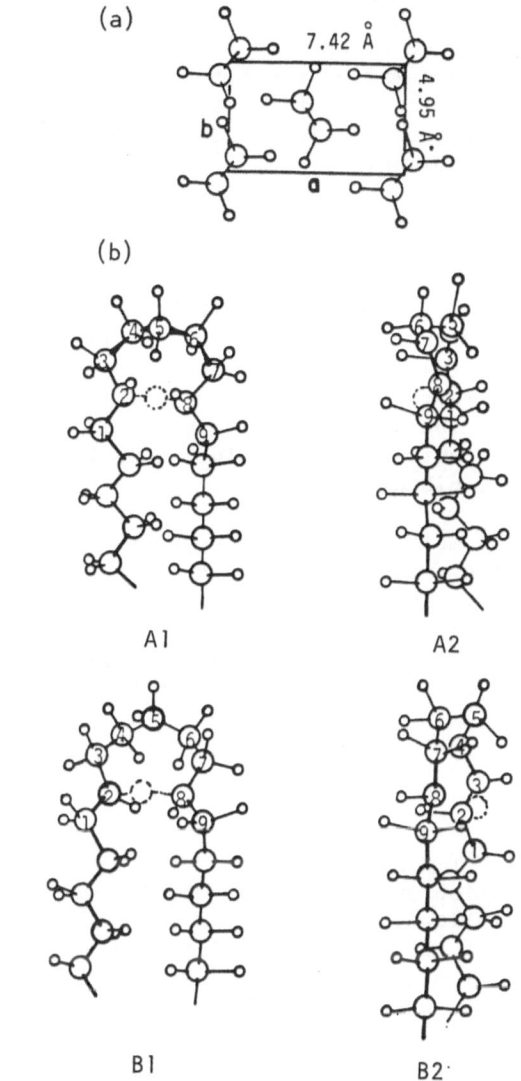

(a)

7.42 Å

b

4.95 Å

a

(b)

A1   A2

B1   B2

Figure 12.   Models of polyethylene chain folding.

     The temperature dependency of the peak intensities indicates
that significant conformational changes occur at the surfaces of
single crystals in the temperature range $30°-80°C$. These temperature effects coincide with the dynamic mechanical α process. It
has been suggested that this relaxation involves reorientation of
the surface folds (15). By demonstrating the rotational mobility
of chain segments in the surface in the temperature range $30°-80°C$,
the infrared results provide direct evidence for the proposed α
relaxation mechanism.

The level of bromination used in this study is close to surface saturation (4). Under these conditions it is likely that more than one carbon atom in the fold may be substituted. Steric considerations indicate that the second substitution is likely to occur on an adjacent carbon atom, for example on carbons 4 and 5 or 5 and 6, to produce a vicinal dibromide.

Many of the vicinal dibromide bands overlap secondary C-Br bands, however, a band at 594 cm^{-1} in the spectra of single crystals cannot be assigned to a secondary C-Br mode. This band was observed previously in brominated single crystals (6) but is not observed in randomly brominated polyethylene nor in crystals grown from randomly brominated polyethylene (9). Since the band does not disappear when the single crystal structure is destroyed either by melting or dissolution it cannot be assigned to a secondary C-Br mode with an unusual conformation unique to the fold. This band may indicate the presence of vicinal dibromides.

Examination of the molecular models suggests that substitution at the most exposed sites is likely to produce meso dibromides in the gauche conformation. This conformation is associated with a band at 600 cm^{-1} in the model compounds. The higher intensity of the 594 cm^{-1} band in diamond-shaped crystals, compared to truncated crystals, may indicate that the folds are tighter and conform more closely to the model.

CONCLUSIONS

Infrared analysis of single crystals specifically labeled at the surface by bromination and of various model compounds has led to the following conclusions:
1. The fold segments of brominated dendritic and diamond-shaped crystals contain all six possible conformations of adjacent C-C bonds. Of the six, the TG' conformation is the most probable.
2. Changes in conformational populations are observed between 30°-80°C in the temperature regime of the dynamic mechanical α process.
3. A C-Br stretching band at 694 cm^{-1} is attributed to meso vicinal dibromides in the gauche conformation.
4. Similarities in the infrared spectra of bromocyclooctane and brominated crystals have led to .chain fold models which consist of nine carbon atoms with the crown or twisted-crown structures.

REFERENCES

1. A. Keller, Philos. Mag., 2, 1171 (1957).
2. E.W. Fischer, Z. Naturforsch. Tel A, 12, 753 (1957).

3. P.H. Geil, J. Polym. Sci., 17, 447 (1957).
4. I.R. Harrison and E. Baer, J. Polym. Sci., A-2, 9, 1305 (1971).
5. M. Eguiluz, H. Ishida and A. Hiltner, J. Polym. Sci. Polym. Phys. Ed., 17, 893 (1979).
6. M. Eguiluz, H. Ishida and A. Hiltner, J. Polym. Sci. Polym. Phys. Ed., 18, 2295 (1980).
7. H.C. Brown and K. Murray, J. Am. Chem. Soc., 81, 4108 (1959).
8. D.J. Blundell and A. Keller, J. Macromol. Sci. Phys., B2, 337 (1968).
9. Y.A. Chang, Ph.D. Thesis, Case Western Reserve University, January, 1982.
10. P. Klaeboe, J.J. Kothe and K. Lunde, Acta Chem. Scan., 10, 1465 (1956).
11. E.L. Eliel, Conformational Analysis, John Wiley, N.Y., 1965.
12. R.G. Snyder, J. Mol. Spectrochim., 28, 273 (1968).
13. B. Wunderlich, Macromolecular Physics, Vol. I, p. 96, Academic Press, N.Y., 1973.
14. P. Corradini, V. Petraconne and G. Allegra, J. Macromol. Sci. Phys., B4, 770 (1971).
15. J.D. Hoffman, C. Williams and E. Passaglia, J. Polym. Sci., C14, 173 (1966).

SUBSTITUTION REACTIONS ON HALOGENATED POLYETHYLENE

Ian R. Harrison, Jacqueline S. Butler, and J. P. Runt

Polymer Science Section
Material Science & Engineering Department
325D Steidle Building
The Pennsylvania State University
University Park, PA 16802

ABSTRACT

It has been previously demonstrated that halogenation of polyethylene single crystals can be selective. Bromination takes place
primarily in the amorphous regions without destruction of the crystalline core. These substituted crystals show unique melting and annealing behavior. Given that a Br substituent has such a marked effect on properties it was of interest to examine the effects of a much
larger substituent. This was accomplished by replacing Br with a
toluene group via a Friedel Crafts reaction. Neutron activation
analysis, FTIR and thermal analysis were used to examine the reaction
products. These techniques show that Br had indeed been replaced by
toluene. Further substitution is primarily in the 1,4 (para) position with some 1,3 (meta) substitution. The presence of bulky substituents modifies the melting behavior of the lamellae. However the
heat of fusion remains unchanged. The major effect on the melting
process is the suppression of any reorganizational annealing behavior.

INTRODUCTION

The nature of the fold surface of polyethylene (PE) single crystals can be studied by means of chemical reactions. Both chlorination
and bromination studies have been carried out on PE lamellae in order
to gain a better understanding of the surface structure [1-3]. It
has been shown that selective halogenation of PE single crystals occurs only in the amorphous regions without destruction of the crystalline core [1,2]. Less than 2% of the bromine uptake is due to

75

reaction with terminal vinyl bonds; the majority being due to substitution at the amorphous - CH_2 - units [1,2].

Studies carried out on the annealing and recrystallization behavior of brominated crystals demonstrates the rather unique influence
which a substituent has on these materials. For example, brominated
sites (CHBr units) act as defects along the crystal surface [1].
Although the defects do not disrupt the crystalline portion of the
single crystal, they greatly reduce the crystallinity of the melt
crystallized samples as seen by a decrease in heat of fusion (c.f.
unbrominated PE). Annealing studies have shown that reorganization of
the crystal is reduced and ultimately eliminated with increasing
bromine content.

Given that a Br substituent has such a marked effect on properties it was of interest to examine the effect of much larger substituents. Substitution studies such as these have been carried out
using polyvinyl chloride (PVC). For example, Teyssie and Smets successfully substituted large aromatic groups (benzene, toluene, etc.)
for the chlorine atoms along a PVC chain in solution [3].

In contrast to PVC in solution, we have studied the substitution
of large-groups on brominated PE single crystals in suspension. Preliminary studies indicate that benzene, toluene and m-xylene can be
substituted on PE crystal surfaces. At this time we will report in
detail the toluene substitution on PE crystals. Towards this end,
the brominated surface of PE crystals were used to introduce toluene
by means of a Friedel-Crafts reaction.

EXPERIMENTAL

Bromination

Single crystals of Marlex 6001 PE were grown from a 0.1% xylene
solution at 87° (Tc) using a self-nucleation technique [5]. After
washing several times with fresh xylene heated to Tc, the suspension
was cooled and transferred to distilled CCl_4. The crystal suspension was placed in a reaction vessel heated to 30°C under a constant
flow of nitrogen. Elemental bromine was added in the ratio: 1 gm
crystals/240 ml CCl_4/6 ml Br_2. After illuminating the mixture with
ultraviolet light for a specified time, samples were withdrawn and
washed with CCl_4 in order to remove excess bromine. The crystals
were then exchanged to acetone to remove CCl_4 and any residual bromine.

Friedel-Crafts Reaction

Seventy-five percent of the remaining crystals were washed with
distilled toluene and placed in a reaction vessel maintained at 35°C
under a constant flow of nitrogen. A small amount of $AlCl_3$ was added

to initiate the reaction. After 1-1/2 hours in suspension, the
crystals were washed with methanol (to complex the initiator) followed
by several washings with CCl_4. Samples were then dried under vacuum
(5 x 10^{-5} mm Hg) for several days.

Characterization

The heats of fusion and melting points of the dried crystals were
obtained using a Perkin Elmer DSC 2. Samples weights ranged from
0.08 mg to 0.2 mg and were weighed to 1% accuracy. The heating rate
was 20°K/minute. The instrument was calibrated using the melting
point and heat of fusion of a known weight of indium. In order to
express the heat of fusion of the PE crystal alone, the bromine and
toluene contents were subtracted from the sample weight and replaced
by the equivalent weight of hydrogen. The reported values were the
average of at least 2 runs. The melting point of the sample was
taken as the point of intersection of the leading indium edge drawn
through the PE sample peak baseline [6].

Samples for Fourier transform infrared spectrsocopy (FTIR) were
prepared by pressing a given amount of KBr-crystal mixture into a disk
at a load of 10,000 lbs. for 5 minutes. The infrared spectra were
recorded and stored by a Fourier transform infrared spectrometer
(Digilab FTS-15/B). A high signal-to-noise ratio was obtained by
taking a minimum of 400 scans of both the specimen and reference at
a resolution of 2 cm^{-1}.

Samples taken from both states (bromination and Friedel-Crafts)
were tested for bromine and chlorine content by neutron activation
analysis. The extent of substitution was determined by the relative
amounts of bromine content before and after the Friedel-Crafts (FC)
reaction.

RESULTS AND DISCUSSION

The substituted products (Br and FC) were analyzed for bromine
and chlorine content. Analysis by neutron activation shows that as
a function of time, the bromination reaction substitutes varying
amounts of bromine on the sample. Sample I was brominated for 6 hrs.
while Sample II was brominated for 20 hrs. Little or no chlorine
(Cl) was detected. Neutron activation analysis values for substituted PE samples (Br and FC) are shown in Table 1. After the FC
reaction, effectively all of the Br is removed and presumably replaced by toluene (95-98% reaction).

Infrared

Three general statements concerning the samples can be made
based on the IR spectra of the PE and FC crystals. Comparison of the

TABLE I

Analysis of Br and FC Samples

Sample	% Br	% Cl + .01%
Orig.		
Br I	1.2%	.01% Cl
Br II	14.3%	.01% Cl
FC (I) (from Br I)	.06%	.02% Cl
FC (II) (from Br II)	.27%	.03% Cl

FC(II) spectra with that of toluene shows that the FC crystals contain no free toluene, i.e. no solvent contamination. Secondly, carbon tetrachloride has strong absorption bands around 1500 and 750 cm^{-1} which are not present in the brominated spectra. (Neutron activation analysis also showed very little chlorine present in all samples.) Thirdly, little or no oxidation of the PE samples occurred, as evidenced by the presence of only very weak carbonyl absorptions in the 1700 cm^{-1} region.

Infrared spectra of the PE and PE substituted crystals from 950–750 cm^{-1} are shown in Figure 1. The most distinctive feature in this region of the spectrum, for the unsubstituted crystals, is a sharp band at 909 cm^{-1}. This is indicative of the C–H wagging vibrations of terminal vinyl groups. This band disappears after bromination due to addition across the terminal double bonds. In fact, it was shown in a separate study that at least 98% of the vinyl end groups react upon addition of bromine implying that practically all chain ends are external to the crystal lattice [6].

After the FC reaction, bands characteristic of aromatic substitution develop in the 900–750 cm^{-1} region as well as at 1607 cm^{-1}. The peak in the FC(II) spectrum at 783 cm^{-1} has been reported to be indicative of 1,3 disubstitution on the ring (meta addition) whereas the more prominent band at 815 cm^{-1} indicates 1,4 disubstitution (para addition) [4]. A peak at 1607 cm^{-1} (not shown) is believed to be due to an aromatic C–C multiple bond stretch. The spectrum of FC(II) is very similar to that of FC(I); the major differences being that the bands in FC(II), due to the aromatic species, have a larger relative intensity than those in FC(I).

The 700–450 cm^{-1} region corresponds to the C–Br stretching modes (Fig. 2). Characteristic C–Br bands are exhibited in this region which are absent in the unbrominated samples. Relatively strong

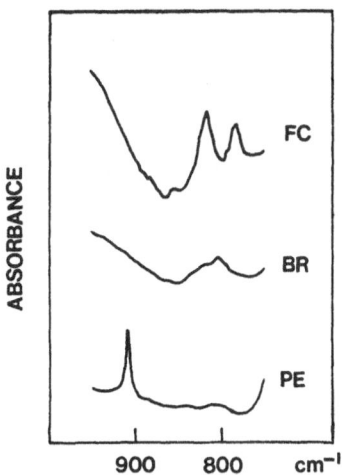

Figure 1:   FTIR absorbance in the region 950-750 cm^{-1} for the original crystals (PE); the same crystals after bromination
(Br) and also after Friedel-Crafts reaction, FC(II).

absorption peaks at 648, 570 and 550 cm^{-1} are seen with weaker absorption peaks at 618 and 525 cm^{-1}. It has been suggested that the
peaks at 648 and 570 cm^{-1} are due to the C-Br stretching modes of
terminal -CHBr - CH$_2$Br groups produced by bromine attack across
terminal vinyl bonds, while the band at 500 cm^{-1} was assigned to C-Br
stretching of secondary carbons [2]. The weaker peaks at 618 and
525 cm^{-1} have been assigned to trans and gauche conformations of
bromine along the terminal vinyl groups [7].

Bands characteristic of aromatic substitution develop in this
region after the FC reaction. Absorption peaks at 653 and 512 cm^{-1}
are indicative of 1,4 disubstitution on the ring while the peaks at
573 and 477 cm^{-1} are indicative of 1,3 disubstitution [8].

A broad absorption peak centered around 1160 cm^{-1} is also seen in
the FC(II) spectra (not shown). There are a number of possible
causes for this absorption. For example, defects in the crystalline
structure upon substitution, absorbance peaks for meta and para
toluene substitution and possible water contamination of the sample
could all contribute to this broadening.

Thermal

In order to determine the effect of substitution on the melting
behavior and crystallinity of PE lamella, both heats of fusion and
melting points were measured. Typical values are shown in Table II.

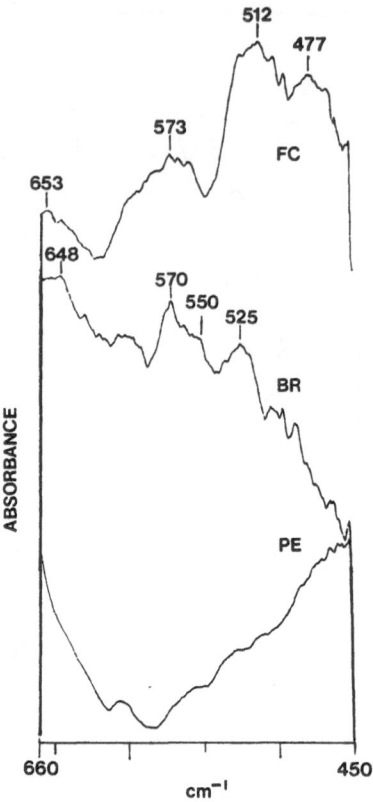

Figure 2: FTIR absorbance in the region of 660–450 cm^{-1} for
           original crystals (PE), the same crystals after bromina-
           tion (Br) and also after Friedel-Crafts reaction FC(II).

     From Table II, one can see that the heats of fusion of the
single crystals are apparently unchanged on substitution (Br and
FC). The interpretation of the constant value is that the crystalline portion of the single crystal is largely unperturbed by
substitution (Br and FC).

     DSC endotherms for both brominated and FC(II) crystals are shown
in Figure 3, however, peak positions have not been corrected for
indium and are only used to show relative positions. (The corrected
melting data are reported in Table II.) The presence of multiple
melting peaks in PE and the reduction in melting point of the PE
crystals after surface halogenation has been previously observed
and rationalized [1,9]. The melting point increase of the substituted (FC) sample relative to that of the brominated sample can be
rationalized in two ways. By using the equation for equilibrium
melting ($Tm^\circ = \Delta H_f / \Delta S$), an increase in the melting point of the FC

TABLE II

Heat of Fusion and Melting Data for Br(II)
and FC(II) Samples.

Sample	$\Delta H(cal/gm)$	$T_m(°K)$
Original PE Crystals	54.5 ± 1	397.3 ± 0.5
		399.3 ± 0.5
		402.8 ± 0.5
Brominated Crystals (14.3% Br)	54.9 ± 1	395.6 ± 0.5
Substituted (FC) Crystals (14.1% Toluene)	55.2 ± 1	399.0 ± 0.5

sample indicates that the entropy difference on melting has decreased relative to the Br sample. (ΔH_f has remained constant.) One
could argue that the bulkier toluene substituent leads to an increase
in chain stiffness (less mobility) thereby reducing the ΔS term of
system.   Alternatively, one can analyze the increase in melting

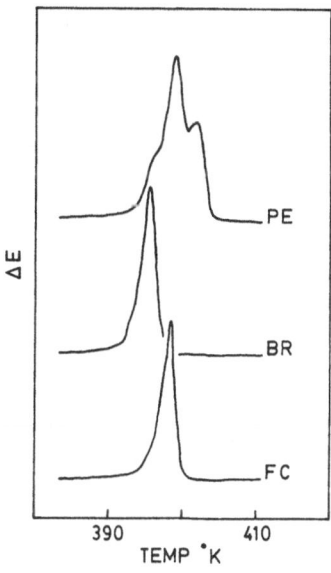

Figure 3:   DSC scans for original crystals, PE; brominated crystals
Br(II); and Friedel–Crafts reacted crystals FC(II).

point of the FC crystals using the Hoffman-Weeks equation (10)

$$Tm = Tm° (1-2\sigma_e/\Delta H_f \ell)$$

Here Tm and Tm° are the melting points of a folded chain and extended chain crystal respectively, σ_e is the surface energy and ℓ is the lamella thickness. Since ΔH and ℓ remain effectively constant throughout the substitution reactions, an increase in Tm of the FC sample implies a decrease in surface energy relative to the brominated sample. The decrease in surface energy can be rationalized by the fact that the toluene substituted surface is similar in character to the polyethylene crystal surface (hydrocarbon). Upon substitution of the toluene groups, the surface of the PE crystal is correspondingly changed from a high energy brominated surface to a lower energy hydrocarbon surface (toluene substituted PE).

The presence of a bulky group on the surface (Br or FC) inhibits any reorganization/annealing processes. As bromine is attached to the surface, lamellar thickening is inhibited since the chains are no longer free to be pulled through the crystal. Annealing studies on substituted crystals have shown that low levels of bromine can be incorporated within the crystal but cause a decrease in ΔH_f (destruction of the crystalline core) [11]. The reduction in the ability to anneal leads to relatively simple one peak thermograms for the substituted samples [12,13].

CONCLUSIONS

Neutron activation analysis, FTIR and thermal analysis show that bromine on the surface of PE single crystals can be replaced by toluene by means of Friedel-Crafts reaction. Infrared analysis of the substituted (FC) cyrstals also shows that the majority of addition occurs by 1,3 and 1,4 substitution to the aromatic ring. The presence of bulky groups on the surface (Br or FC) inhibits any reorganization annealing by the crystals.

ACKNOWLEDGMENTS

The authors would like to thank the NSF (Materials Division) Polymers Program for partial support for this work under grant no. DMR-7825233. We would also like to thank Dale Raupach of the Pennsylvania State University's Breazeale Nuclear Reactor Facility for performing the neutron activation analysis.

REFERENCES

1. I. R. Harrison and E. Baer, J. Polymer Sci., A-2, 9, 1305 (1971).

2. N. A. Narroyo and A. Hiltner, J. Appl. Poly. Sci., <u>23</u>, 1473
    (1979).

3. J. Guzman, G. G. Fatou and J. M. Pesena, Makromol. Chem., <u>181</u>,
    1051 (1980).

4. P. H. Teyessie and G. Smets, J. Polymer Sci., <u>20</u>, 362 (1956).

5. D. J. Blundell and A. Keller, J. Macromol. Sci., Phys., <u>B2</u>, 337
    (1968).

6. Perkin-Elmer Differential Scanning Calorimetry Model DSC-2
    Instruction Manual, Section 403, Norfolk, Conn., 1974.

7. M. Equiluz, H. Ishida and A. Hiltner, J. Polymer Sci., Phys. Ed.,
    <u>17</u>, 894 (1979).

8. F. F. Bentley, L. D. Smithson and A. L. Rozek, <u>Infrared Spectra</u>
    <u>and Characteristic Frequencies</u> ~ 700-300 cm^{-1}, <u>1</u>, John Wiley
    & Sons, New York, 1968, p. 68.

9. I. R. Harrison, J. Polymer Sci., Poly. Phys. Ed., <u>11</u>, 1002 (1973).

10. J. D. Hoffman and J. J. Weeks, J. Res. N.B.S., <u>66A</u>, 13 (1962).

11. I. R. Harrison and E. Baer, Analytical Cal., Vol. 2, R.S. Porter
    and J. H. Johnson, Eds., Plenum Press, N.Y., 1970, p. 38.

12. P. J. Lemstra, A. J. Schonten and G. Challa, J. Polymer Sci.,
    Poly. Phys. Ed., <u>10</u>, 2301 (1972).

13. H. E. Bair and R. Salovey, J. Macromol. Sci., <u>B3</u>, 3 (1969).

STRUCTURE DETERMINATION OF CROSSPOLYMERIZED

POLY(1,11-DODECADIYNE)

M. Thakur and Jerome B. Lando

Department of Macromolecular Science
Case Western Reserve University
Cleveland, Ohio 44106

ABSTRACT

Electron diffraction patterns were obtained from cross-polymerized crystals of poly(1,11-dodecadiyne), formed by casting thin (>200Å) films of uncrosspolymerized polymer from chloroform. Two orientations of these crystals were obtained by varying the evaporation rate of the chloroform solvent. Crosspolymerization resulted from subsequent exposure of these samples to Co^{60} γ-radiation. Thirty-six independent reflections were obtained from the a*c* reciprocal lattice net and a lattice net have b* and the (h 0 $\bar{h}$) reciprocal lattice direction as orthogonal axes. Refinement of the structure was accomplished with these data. The unit cell of the crosspolymerized material was monoclinic, space group $P2_1/n$, a = 9.17Å, b (hydrocarbon chain axis and unique axis) = 12.25Å, c (polydiacetylene axis) = 9.92Å and the angle β = 123.5°.

INTRODUCTION

The crosspolymerization of the macromonomer poly(1,11-dodeca-diyne), using UV, x-ray or Co^{60} γ-radiation has recently been reported[1]. The term macromonomer is used to describe the original polymer which has a chemical repeat unit $(CH_2)_8$-C≡C-C≡C-. The term crosspolymerization is utilized to distinguish systematic polymerization of the diacedylene units to a crystalline structure composed of sheets, as indicated in Fig. 1, from the more familiar random crosslinking that many polymers undergo when exposed to radiation. The crystal structure of the crosspolymerized material was refined using electron diffraction data because of the limited information obtained from x-ray fiber patterns.

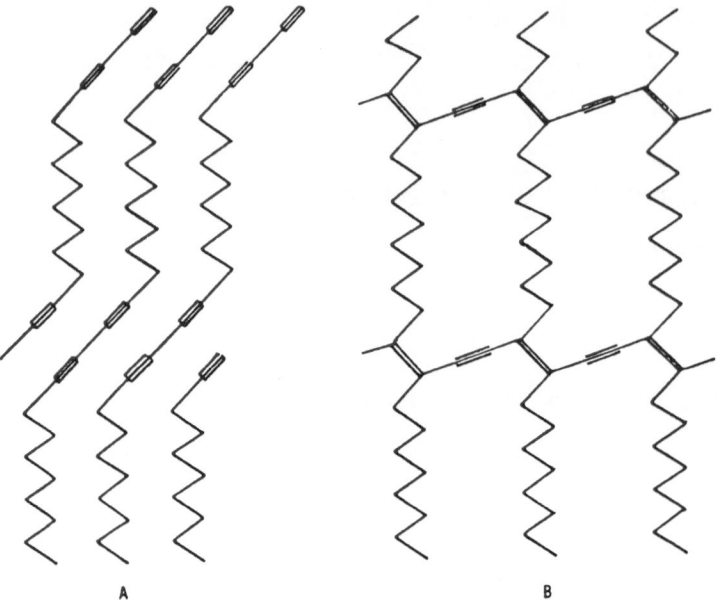

Figure 1. Model of the Reaction of the Macromonomer. A. Macro-
          monomer. B. Crosspolymerized.

EXPERIMENTAL SECTION

      The macromonomer was prepared by oxidative coupling of HC≡C-
$(CH_2)_8$-C≡CH using copper-pyridiene catalyst[1,2,3]. After purification the polymer was dissolved in chloroform to make a dilute solution. A drop of this solution at 4°C, on a carbon coated copper
grid, was evaporated to obtain single crystals. These crystals were
then irradiated by γ-radiation (100 M rad) for complete crosspolymerization. Diffraction patterns were obtained by a JEOL JEM
100 B electron microscope at 100 KV. A second orientation of the
crystals was obtained by casting a film at room temperature. Electron diffraction data were obtained at low beam intensity. The
intensity data were collected using a high precision photodensitome-
ter. The peaks in the intensity spectra were quite sharp; therefore
the peak heights were taken as the relative intensities of the
diffraction maxima. The structure was refined using the "Lals
Mark Six" program originally developed by Arnott and co-workers[4].

      Measurements of the thickness of the crystals were performed
in the following way. The crystals on the carbon coated copper
grid were carbonshadowed at an angle of 45°. Some part of the
substrate remains unexposed to this shadowing due to the height of
the crystal which blocks the carbon particles. The unexposed strip

Figure 2. Electron Diffraction Pattern (a*c* reciprocal lattice net).

manifests itself as a light strip at the edge of the crystal on the
electron micrograph. Since the shadowing angle is 45° the width
of this strip is the thickness of the crystal.

RESULTS AND DISCUSSION

     The diffraction pattern of the a*c* reciprocal lattice net
remained sharp for about 20 secs but that of b*(h0h̄) net (rapid
avaporation) remained reasonably sharp for only about 13 secs even
under the lowest possible beam-intensity. In spite of this we
 tained a reasonable number of reflections by developing the pic-
 res appropriately. Initially only the a*c* data were used for
 finement of the structure. Afterwards the full data were utilized.
  equatorial h0h̄ reflections, which appeared on both patterns,
 are used for scaling. The a*c* pattern showed a c-axis repeat
distance of 9.92Å which is approximately twice the ideal polydiacetylene chain repeat. Therefore one can safely assume that the
diacetylene chain is along the c-axis. The (h0h̄) pattern as well
as x-ray studies gave a b-axis of 12.25Å which obviously fits with
the hydrocarbon chain repeat. This axis is orthogonal to the ac
plane indicating possible chain folding in the macromonomer. Therefore the unit cell is monoclinic. There is a screw axis along b as
demonstrated by the absence of 0k0 k – odd reflections. The h0ℓ

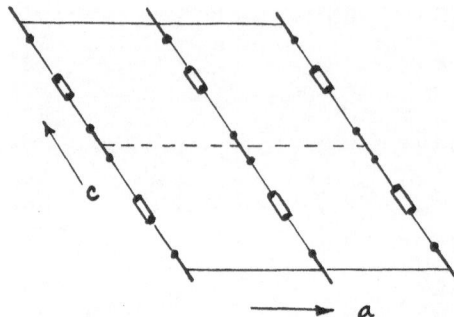

Figure 3.   Crosspolymerized Poly(1,11-Dodecadiyne), Model
            Structure.

plane contains additional systematic absences, h+ℓ = odd, which
indicates a glide plane perpendicular to the b-axis. With this information the space group assigned to this crystal is P2$_1$/n.

   With the above information at hand we can conveniently choose
our starting model as in Fig. 3. We place the diacelylene chain
along the c-axis and the hydrocarbon chain along the b-axis. They
are four repeat units per unit cell and an n-glide in the ac plane
Therefore there should be one chain at the center of the unit cell
A series of models, which were small modifications of our preliminary model, were tried. Initially all these models were tested
against contact calculations (van der Waals interaction). All possible combinations of the four chains with regard to their senses
(UP or DOWN) were taken into consideration. Orientation of the diacelylene rod with respect to the c-c planer zigzag was varied
until contacts were minimized. Both carbon and hydrogen atoms we
used for structure factor calculation. Since the measured thickr
of the crystals is less than 200Å, dynamical effects were ignore
Or to be more precise. the contribution of the dynamical effect
considered to be within experimental error. The existence of
matic absences was a strong indication that this was a reasona
assumption.

   After exploiting the appropriate options provided by Lals we
reached a residual of 0.13. The residual we refined is defined as

$$R = \left\{ \sum_{m=1}^{N} W_m \Delta F_m^2 / \sum_{m=1}^{N} W_m F_m^2 (obs) \right\}^{1/2}$$

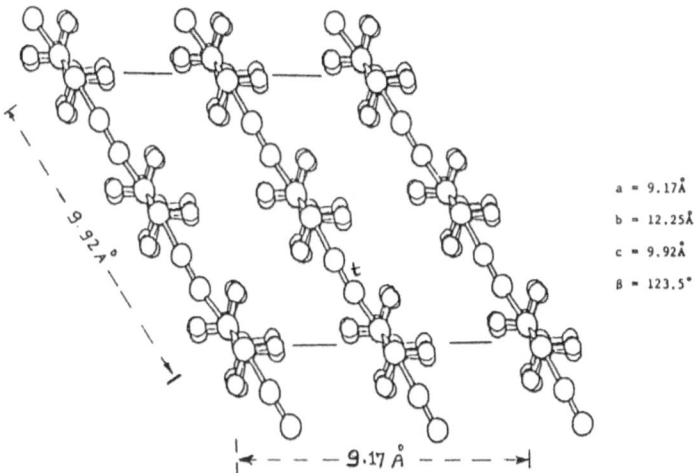

a = 9.17Å

b = 12.25Å

c = 9.92Å

β = 123.5°

Figure 4. ac Projection of Final Structure, t, triple bond.

where $\Delta F_m = |F_m(\text{obs}) - F_m(\text{cal c})|.$

A comparison of calculated and observed structure factors is given in Table 1.

We kept $W_m = 1$ for all diffraction maxima. The projection on the ac plane (Fig. 4) shows that the diacetylene chains remain in the plane of c-c zigzag, which should be energetically favorable. Two UP and two DOWN chains were found to yield the best residual. A DOWN chain is formed by rotating an UP chain by 180° about the a-axis. The ac projection of the DOWN chain is almost identical with that of the UP chain, although the two chains are not identical the projection, Figure 5). This similarity might mislead one to pink that the unit cell should be half of what is given.

b Final bond angles, dihedral angles, and the bond lengths of single, double and triple bonds are given in Table 2. The of atoms in Table 2 are identical to those in Figure 5.

Single crystal electron diffraction patterns of the predomin-day uncrosslinked macromonomer were obtained using high speed may film under very low beam intensity. The unit cell (Table 3) is monoclinic and larger than that of the crosspolymerized polymer. Macroscopically this kind of contraction was observed when the un-reacted macromonomer was crosspolymerized. This kind of behavior has been observed for some other diacetylene polymerizations as well[5]. The space group symmetry is unaltered by crosspolymeriza-tion. Thus the starting model for refining the macromonomer should

TABLE 1

h	k	l	F − CALC	F − OBS
2	0	0	8099	7386
4	0	0	2369	2161
0	2	0	1532	1587
0	4	0	1202	1487
0	6	0	865	1143
0	8	0	682	887
0	0	2	5109	5188
0	0	4	817	1062
1	0	−1	3031	3357
1	0	1	2575	2734
−1	0	3	3458	3479
−1	0	5	526	671
1	0	3	2357	2614
3	0	1	1052	1318
2	0	2	2435	2136
2	0	−2	5329	5140
−2	0	4	949	1294
1	2	−1	572	763
1	4	−1	493	366
1	6	−1	440	898
3	1	−3	271	234
1	1	−1	249	122
2	1	−2	814	244
−1	2	1	293	548
2	2	−2	539	601
−2	2	2	1063	1221
3	2	−3	976	856
4	2	−4	257	671
4	0	−2	2431	2734
4	0	−4	1091	1245
3	0	−3	2864	3113
5	0	−5	1554	1659
5	0	−3	1293	1525
−3	0	5	943	1318
3	0	−1	1430	1807
6	0	−4	711	793

TABLE 2

Atoms Constructing the angle			Bond Angle	Dihedral Angle
C_0	C_1	C_2	110.5	
C_1	C_2	C_3	109.6	179.3
C_2	C_3	C_4	108.8	180.0
C_3	C_4	C_5	110.5	180.0
C_4	C_5	C_6	112.3	180.0
C_5	C_6	C_7	112.5	180.0
C_6	C_7	C_8	112.5	180.0
C_7	C_8	C_9	108.5	180.0
C_8	C_9	C_{10}	112.0	180.0
C_3	C_4	C_{11}	124.0	356.1
C_4	C_{11}	C_{12}	180.0	358.2
C_4	C_5	H_{5A}	110.0	60.0
C_4	C_5	H_{5B}	112.1	302.5
C_5	C_6	H_{6A}	107.4	60.9
C_5	C_6	H_{6B}	106.6	300.0
C_6	C_7	H_{7A}	105.9	60.0
C_6	C_7	H_{7B}	106.4	300.0
C_7	C_8	H_{8A}	113.6	60.0
C_7	C_8	H_{8B}	113.2	301.2
C_8	C_9	H_{9A}	106.1	61.2
C_8	C_9	H_{9B}	106.5	300.0
C_9	C_{10}	H_{10A}	104.5	60.0
C_9	C_{10}	H_{10B}	106.8	300.0
C_3	C_2	H_{2A}	105.3	61.4
C_3	C_2	H_{2B}	107.9	298.5
C_2	C_1	H_{1A}	110.8	60.0
C_2	C_1	H_{1B}	111.2	300.0
C_1	C_0	H_{0A}	106.4	60.0
C_1	C_0	H_{0B}	107.6	300.0

	C – C	C = C	C≡C	C – H
Bond length (Å) (Not refined)	1.534	1.33	1.20	1.065

TABLE 3

Cell Parameters	Macromonomer (unreacted)	Crosspolymerized Material
a	13.25Å	9.17Å
b	14.15Å	12.25Å
c	7.63Å	9.92Å
d	118.5°	124.5°

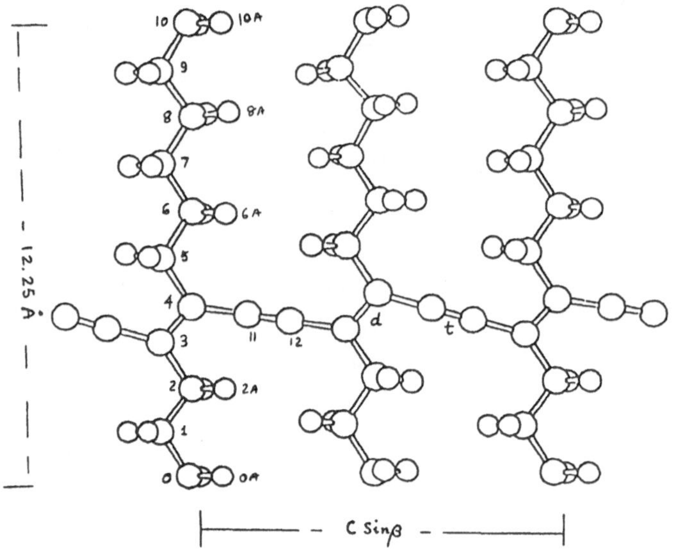

Figure 5.   bc Projection of Final Structure; t, triple bond,
d, double bond.

not be much different from that of the crosspolymerized one.   This
work is now under way.

ACKNOWLEDGMENT

The support of this work by the Office of Naval Research under
Contract N00014-77c-0213 is gratefully acknowledged.

REFERENCES

1.   1. Day, D.R. and Lando, J.B., submitted to Journal of Polymer
      Sci., Polymer letters.
2.   Campbell, I. and Eglinton, G., "Organic Synthesis", Vol. 45,
      p. 39.
3.   Campbell, I. and Galbraith, A., J. Amer. Chem. Soc., 182, 889
      (1959).
4.   Campbell, P.J. Smith and Arnott, Struther, Acta Cryst. A34, 3
      (1978).
5.   Day, David R. and Lando, J.B., Macromolecules, 13, 1483 (1980).

STRUCTURE AND MORPHOLOGY OF NASCENT POLYETHYLENES

OBTAINED BY TiCl$_3$ HETEROGENEOUS ZIEGLER-NATTA CATALYST

A. Muñoz-Escalona, C. Villamizar, and P. Frias

Laboratorio de Polímeros, Centro de Química
Instituto Venezolano de Investigaciones
  Científicas (IVIC)
Apartado 1827, Caracas 1010-A, Venezuela

INTRODUCTION

Structural and morphological studies of the as-polymerized polyolefins have received considerable attention in the voluminous literature on Ziegler-Natta polymerization, due to scientific and technological reasons (1-3). The first studies concerned the growth of the polymer on the catalyst surface in an attempt to get more information about the very complex polymerization mechanism, by which the olefin is converted into the polymer (4-6). On the other hand, by controlling the crystallization process of the polymer during its formation (simultaneous polymerization and crystallization) the possibility of obtaining extended chain crystals of polyolefins could be considered (7,8). Finally, the studies carried out in the industry are motivated to control the overall morphology of the forming polymer particles inside the reactor, due to the fact that its size, shape and porosity affect the stirrability of the polymer mass, the rate of polymerization and consequently the polymer production (9). In reference to this, the coming third generation of Ziegler-Natta catalysts will be those that, by controlling the overall nascent morphology (average size and distribution, shape, bulk, density, etc.) of the polymer particles, the pelletizing step of the process could be unnecessary using high active catalytic systems (10).

Three principal morphologies (films, globular and worm-like) of nascent polyethylenes obtained by heterogeneous Ziegler-Natta catalysts have been described in the literature (11,12). The cobweb and fibrillary morphologies are essentially due to secondary effects that take place during the polymerization reaction, such as stretching of the growing original morphologies at any stage of the reaction (13).

The formation of the principal morphologies has been related to the
polymerization rate and activity of the catalytic system (14). However, it is not totally clear whether the observed morphologies are
to some extent consequential of the polymerization techniques used
(with and without stirring, presence or absence of solvents, polymerization temperature, etc.) or are exclusively controlled by the
catalyst. For that reason, in this paper, ethylene was polymerized
by different polymerization techniques (gas-phase polymerization,
slurry and solution polymerization) using $TiCl_3$ as catalyst.

EXPERIMENTAL

Polymerization Techniques and Catalytic Systems

     Slurry polymerization of ethylene was carried out in n-
paraffine, under stirring (10-1000 rpm) and below the solution temperature of the forming polymer (0-115°C), so that the polymer crystallizes during polymerization. The catalytic system used was $TiCl_3$
grade AA from Stauffer Chemical Co., and with Et_3Al as co-catalyst.
The polymerization was performed in batch, in a one-liter glass
autoclave from Büchi (Switzerland) at 1-10 atm. ethylene pressure.
Withdrawal of the samples from the reactor at different polymerization times and other polymerization procedures has already been
described in a previous paper (8). Solution polymerization of ethylene was carried out in the same system and conditions but just above
the solution temperature of the polyethylene (120°C), so that the
polymer started to dissolve during the reaction. The ratio of
$|Al|/|Ti|$ used in all experiments was 5:1 except where otherwise
noted.

     In order to find out whether the presence of solvents and also
the application of stirring have any effect on the nascent morphologies, ethylene was polymerized from the gas-phase. Polymerizations
were carried out in a schlenk-type reactor, at 0.3 atm. ethylene
pressure on layers of $TiCl_3/Et_2AlCl$, without solvents and stirring
(quiescent conditions), and a polymerization temperature between 0
and 120°C (gas-phase polymerization). In order to observe, under
the scanning electron microscope (SEM), the growth of the polymer
on the catalyst layer, glass microscope slides were placed on the
bottom of the schlenk. After the reaction was completed the glass
slides supporting the catalyst layers with the polymer, were
removed and treated for SEM examination.

Samples Preparation and Characterization

     In order to observe the polymer growing on the surface of the
catalyst layers and the morphology of the samples, they were carefully deposited on SEM stubs and joined by conductive-adhesion silver
paint. The samples were then coated with carbon under vacuum for

energy-dispersive X-ray analysis or with gold by sputtering technique, for morphological observation. The so prepared probes were
examined at 50 KV with a Phillips SEM 500.

The preparation of the samples for observation under the transmission electron microscope (TEM) was performed in the usual manner.
The TEM grids were covered first with a carbon film. Then, a suspension of the samples in ethanol was treated in an ultrasonic bath
for dispersion. A drop of the suspension was gently placed on the
grids and the solvent evaporated. Finally, the samples were observed
in a Phillips TEM 201 at 100 KV after being shadowed with Pt-C at an
angle of 35° to improve the contrast.

Intrinsic viscosities were measured at 135°C in decalin. The
viscosity average molecular weights were calculated from intrinsic
viscosities using the equation (15):

$$|\eta| = 6.2 \times 10^{-4} \, \bar{M}_v^{0.70} \ (dl/g)$$

RESULTS AND DISCUSSION

Polymerization Rate and Molecular Weights

The amount of PE per weight unit of $TiCl_3$ obtained by the three
polymerization processes are given in Figs. 1 and 2. The comparison
between the various curves show that the polymerization rates obtained for the gas-phase technique at different temperatures (0-120°C)
(Fig. 1) are the lowest and the polymerization rate obtained in solution (120°C in n-paraffine) the highest (Fig. 2). For the slurry
and gas-phase techniques the polymerization rate increases with the
temperature.

On the other hand, due to the fact that the ethylene pressure
has no significant influence on the molecular weight (16), the comparison between the molecular weights of the PE can be made and are
given in Table 1. It can be seen that the molecular weights obtained
in the gas-phase polymerization are the highest. Similarly, for each
polymerization technique, the molecular weights depend strongly on
the polymerization temperature. Therefore, at the same polymerization temperature the molecular weights obtained in the gas-phase
technique are higher than for the slurry. However, the molecular
weights obtained at 120°C in the gas-phase are only slightly higher
than those obtained at the same temperature in solution.

Electron Microscope Observation of the Nascent PE Obtained in the
Slurry and Solution Processes

Fig. 3a shows a general aspect of the overall morphology of
polymer particles produced during the slurry polymerization of

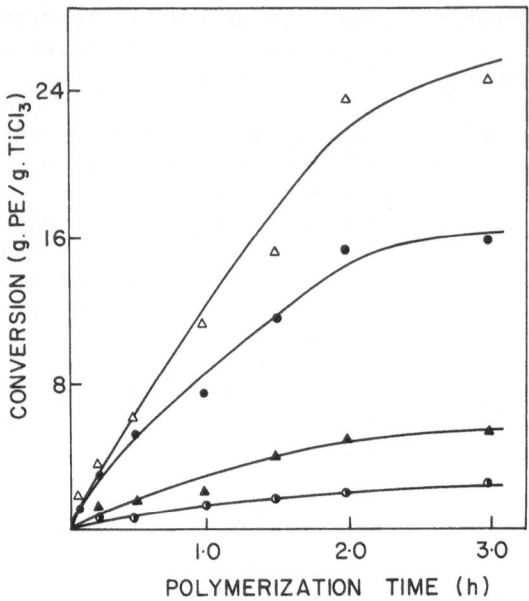

Figure 1. Activity of TiCl$_3$/AlEt$_2$Cl in the gas-phase polymerization
           of ethylene under quiescent conditions, at 0.3 atm. and
           different temperatures: 0°C; Δ 25°C; 0 70°C; 120°C.

ethylene, using the heterogeneous catalytic system TiCl$_3$/Et$_3$Al.
The particles present a broad distribution in size and it has been
suggested that they retain the shape of the catalyst granules by the
so-called replication process (17). Energy-dispersive X-ray analysis
of the elements Ti, Al and Cl took on different single spots of the
particles revealed that the catalyst is dispersed all over.

    At higher magnification different interesting features could
be observed. The particles were usually covered by a stretched
polymer film (Fig. 3b). Under this film different kinds of morphologies (globular and worm-like) were observed depending on the
polymerization temperature or polymerization rate (Fig. 3c). At low
polymerization temperature (0°C) only a globular or nodular morphology was observed. However, at 70°C both types of morphologies
could be observed. It seems to be, that the worm-like morphologies
are assembled by the one directional aggregation of several globules
as shown in Fig. 4. The interconnecting fibers are originated by
separation of these morphologies and deformation or stretching of
the polymer mass between them, due to the expansion of the polymer

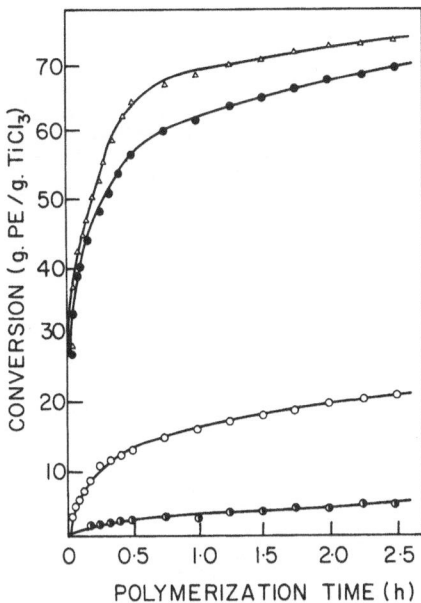

Figure 2. Activity of TiCl$_3$/AlEt$_3$ in the polymerization of ethylene in n-paraffine as reaction medium and different conditions: O in slurry at 0°C, 1 atm and 10 rpm; O in slurry at 0°C, 10 atm. and 1000 rpm; O in slurry at 70°C, 10 atm. and 1000 rpm; Δ in solution at 120°C, 10 atm. and 1000 rpm.

particles during its growth. In order to study the fine detail of the globular and worm-like morphologies, selective oxidation treatments with fuming nitric acid were carried out at 70°C during 32 h. Electron micrograph observations indicated that the globular morphologies are composed by an agglomeration of very small spheres. Furthermore, it appeared that the worm-like morphologies were built up by the aggregation of twisted helicoidal ribbon-like crystals as shown in Fig. 5. Further information about the fine details of the different morphologies found under SEM could be obtained by dispersion of the samples with an ultrasonic bath and examination under TEM. At low polymerization rate a lamellar morphology was generally observed. At high polymerization temperature (>25°C) a ribbon-like crystal could be seen (Fig. 6). The length and amount increases with the polymerization temperature, polymerization rate and by annealing. On the other side, the thickness measured from the length of the shadow ranges between 150-300Å. These values agree very well with the thickness calculated from the DSC melting temperatures (see also Ref. 8). The width of the ribbons is very irregular, ranging between 400-2000Å. Furthermore, it appears that the ribbons

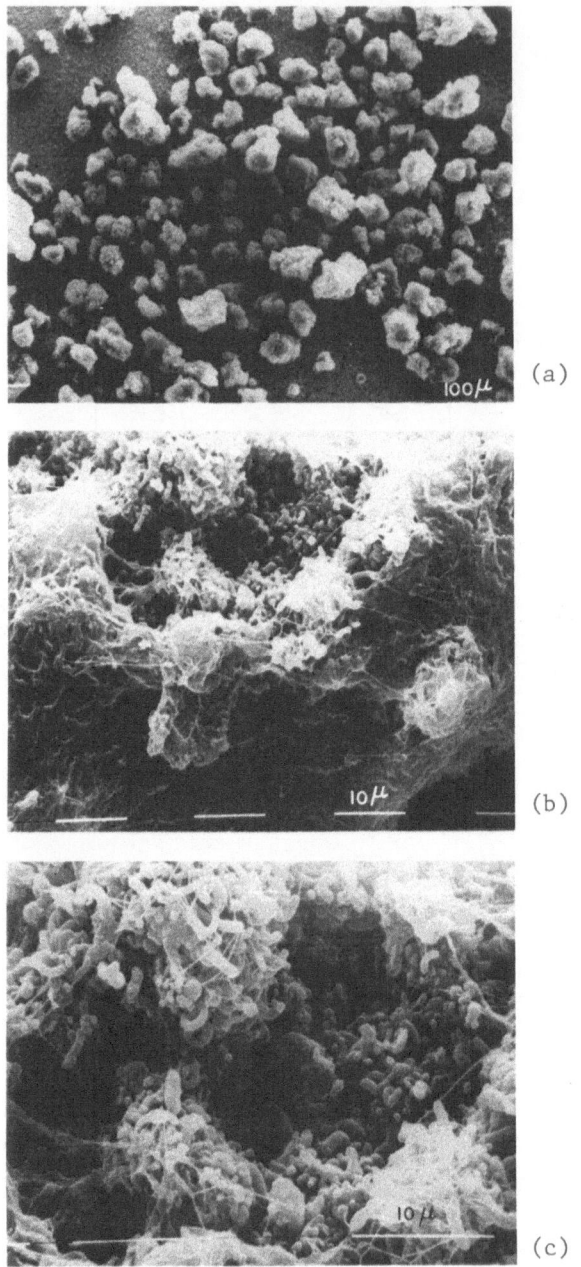

Figure 3.   Scanning electron micrographs of polyethylene particles
            obtained by polymerization in slurry: a, b and c, at
            different magnifications.

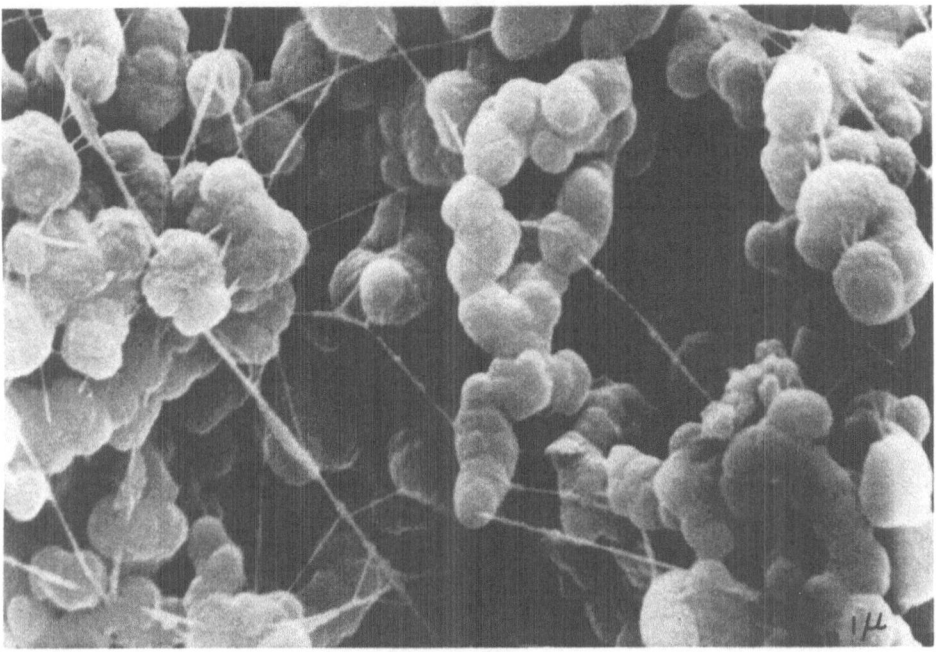

Figure 4. Scanning electron micrograph of as-polymerized polyethylene obtained by polymerization in slurry.

are formed by aggregation of microspheres or platelets, probably
originated by tiny catalyst particles. Depending on the arrangement
of the catalyst particles, lateral ramifications can be formed as
shown in Fig. 7a. Electron diffraction on all lamellar and ribbon-
like crystals revealed that the molecular chain axis is perpendicular to the lamellae surface or to the ribbon axis and they are frequently twined. Therefore, the macroconformation of all crystals is
basically chain folded. Fuming nitric acid oxidation of the ribbon-
like crystals (Fig. 7a) confirms that they are formed by joined
microspheres or platelets of the kind seen in Fig. 7b. This figure
also suggests that the microspheres can be added in different forms
giving rise to all types of morphologies found, such as: lamellae,
globulars and ribbons with and without ramifications.

     Finally, at the beginning of the polymerization in solution
(120°C polymerization temperature) a rather smooth layer of polymer
was observed. But as the process continued (high polymerization
time), a partial formation of worm-like morphologies were still
present indicating that the polymer morphologies were formed first
and then started to dissolve.

Table 1

Average Viscosity Molecular Weight of PE Obtained by the
Three Polymerization Techniques, using $TiCl_3$ as Catalyst

Run N^o	Polymerization temperature (^oC)	Polymerization time (min.)	Molecular Weight $\bar{M}_v$ x 10^{-6}
	Slurry Polymerization[a]		
6-3	0	45	1.06
6-4	0	150	1.46
1-3	70	45	0.60
1-4	70	150	0.78
	Solution Polymerization[a]		
11-2	120	20	0.24
11-3	120	45	0.29
11-4	120	150	0.32
	Vapor Phase Polymerization[b]		
40	70	90	1.15
35	70	120	1.22
46	120	30	0.30
44	120	120	0.45

[a] In n-paraffine as solvent, under stirring at 10 atm. ethylene
pressure, using Et_3Al as co-catalyst.

[b]
  Without solvent or stirring, at 0.3 atm. ethylene pressure, using
Et_2AlCl as co-catalyst.

Electron Microscope Observation of the Nascent PE Obtained in the
Gas-Phase Process

    Gas-phase polymerization carried out in static conditions gave
rise to very similar morphologies. However, under these conditions
the formation and growth of the nascent morphologies could be observed and followed more clearly under the electron microscope. It was
found that at the beginning of the reaction the polymer forms a
film covering the catalyst layers (see Fig. 8a). In addition to
this, dots, microspheres or platelets were observed, probably originated by tiny catalyst particles having very high activity for the
ethylene polymeration and separating themselves from the main bulk.
When the reaction proceeds, globular and worm-like morphologies were
also observed, which are built-up by an aggregation of microspheres
or platelets as shown in Fig. 8b. Finally, TEM observation revealed
the formation of ribbon-like crystals, which grew in length as the

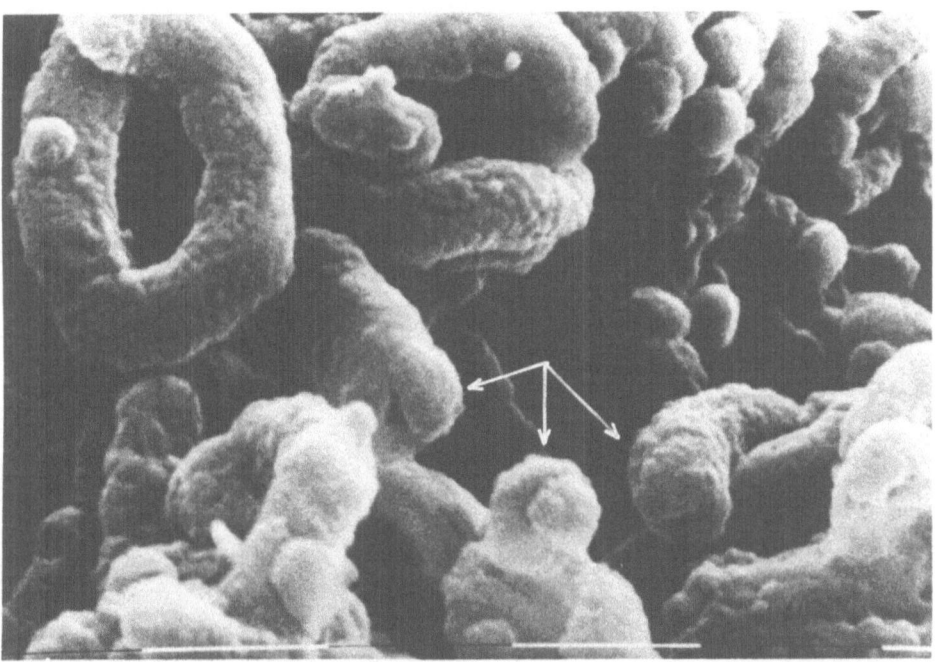

Figure 5. Scanning electron micrograph of as-polymerized polyethyl-
           ene obtained by polymerization in slurry and treated with
           nitric acid at 70°C for 32 h. The arrow shows dtails of
           the worm-like morphology.

polymerization in time and temperature were increased. Similarly,
the ribbon-like crystals were formed by an aggregation of microspheres or platelets as shown in Fig. 8c. Energy dispersive X-ray
analysis carried out under the scanning electron microscope and
taken from single spots of the catalyst surface, revealed different
types of active centers. Those showing high activities, and giving
rise to the dots, ribbon-like and worm-like morphologies were found
to have ratios of Cl:Al:Ti of about 4 to 3:1:1 (Fig. 9). These
ratios correspond very well with the chemical composition of the
bimetallic active complexes (I and II) proposed by Rodriguez et al.
(18) and Natta et al. (19), respectively: (See top of next page).

     On the contrary, analysis taken from parts of the catalyst
having very few polymers, showed chemical compositions with very
low amounts of Al. Finally, no Al at all was detected in areas
without polymer. These results gave experimental evidence supporting
the fact that the presence of the metal alkyl is needed for
formation of the active catalytic complexes.

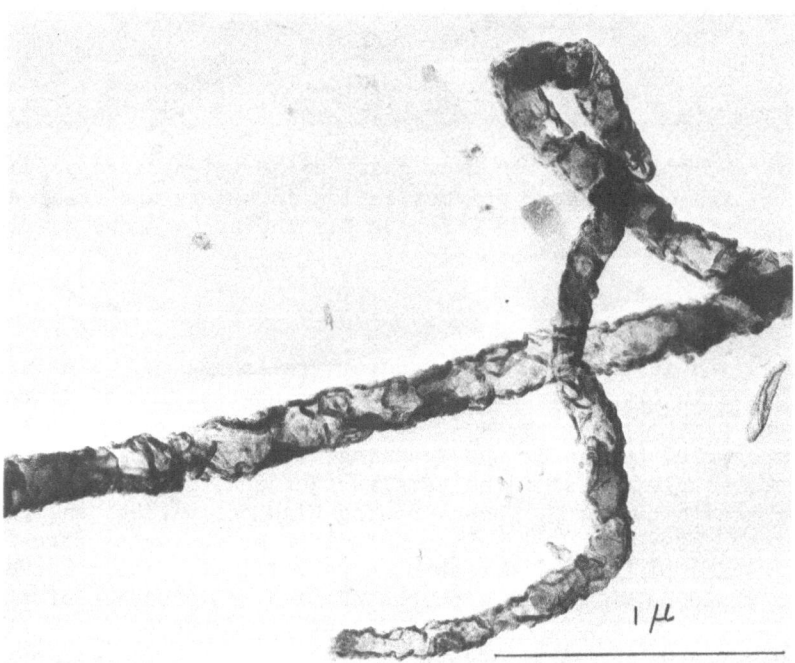

Figure 6.   Transmission electron micrograph of ribbon-like crystals
            obtained by polymerization in slurry of ethylene.

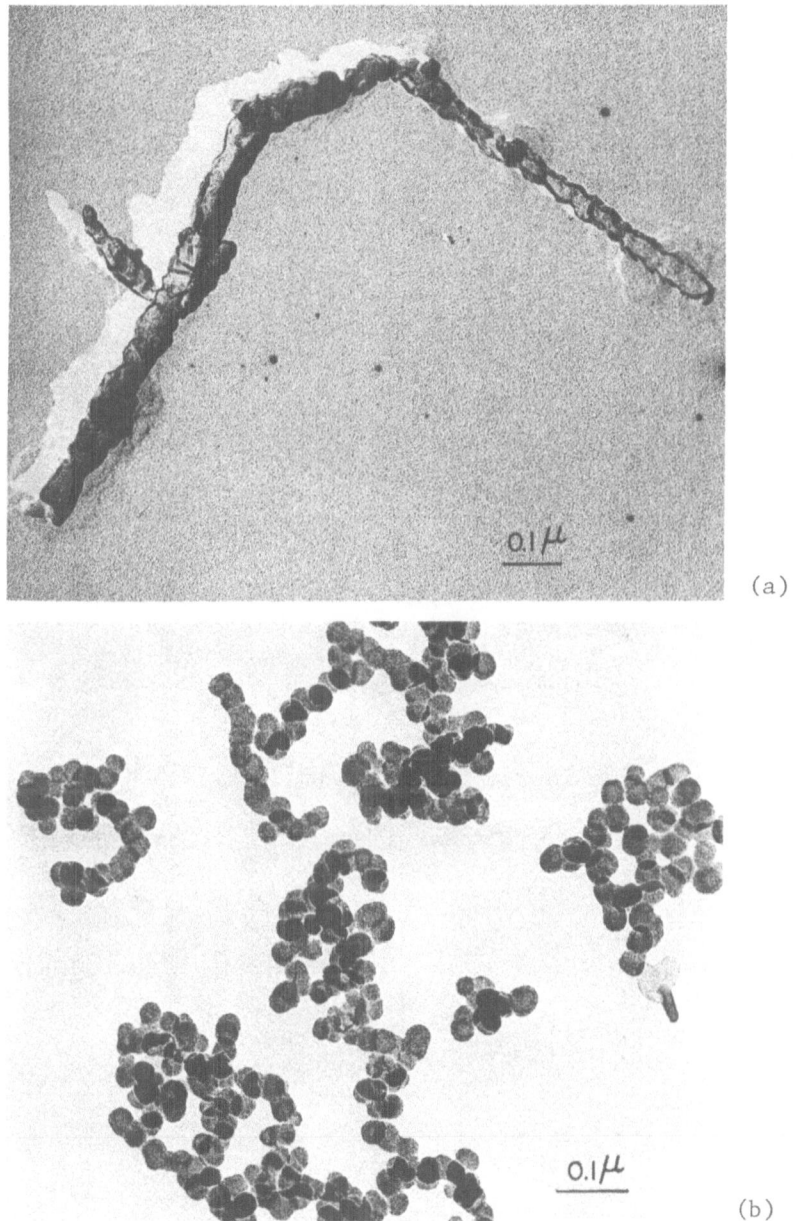

Figure 7. Transmission electron micrograph of ribbon-like crystals
(a) without and (b) with nitric acid treatment.

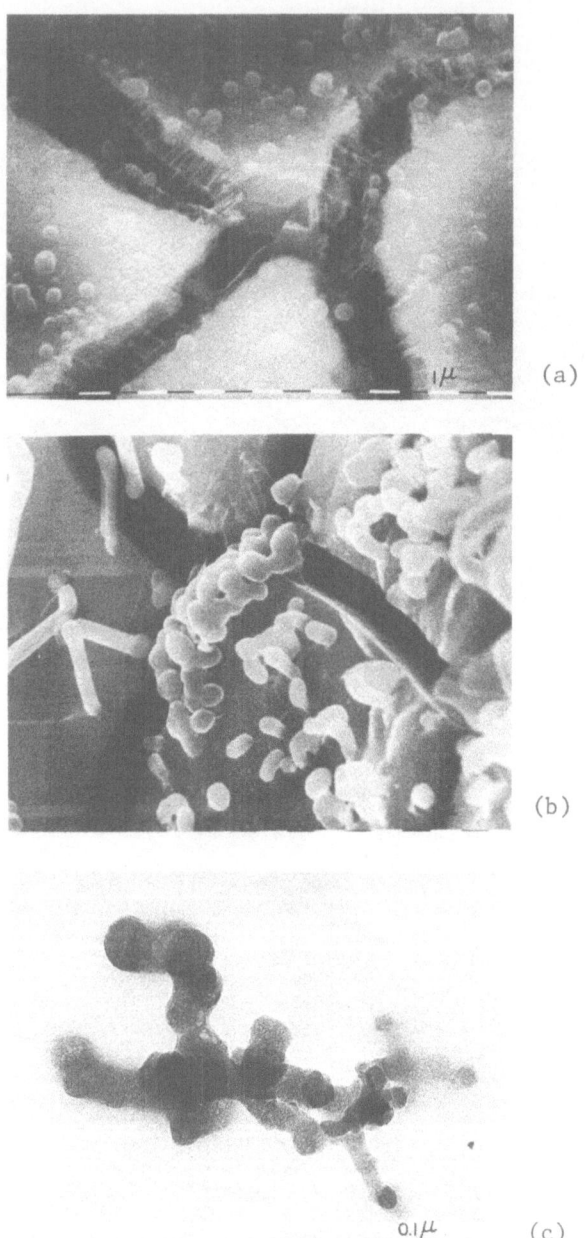

Figure 8. Morphologies of as-polymerized PE obtained by gas-phase
            polymerization: a) and b) scanning electron micrographs
            and c) transmission electron micrograph.

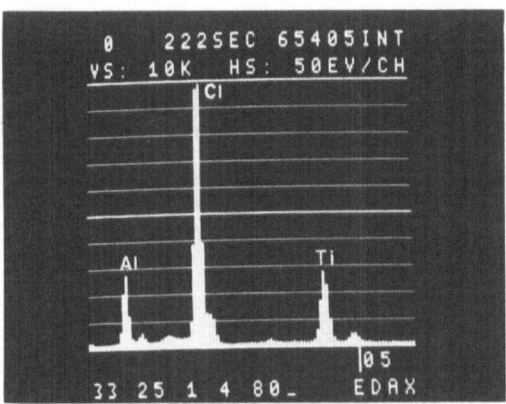

Figure 9.   Energy dispersive X-ray spectrum.

The morphology obtained in the gas-phase polymerization at 120°C reaction temperature was very interesting and consisted of very well defined stacks of lamellae, which maintained the overall appearance of the globular and worm-like morphologies (see Figs. 10a and 10b). By annealing at 130°C the globular and worm-like morphologies obtained in the slurry polymerization did not give rise to lamellae formations. Therefore, the lamellae found in the gas-phase must have been formed during the polymerization. Furthermore, under TEM ribbon-like crystals showing striations were frequently found (Fig. 11a). This morphology is very different from that found in the slurry polymerization (compare Fig. 6 with Fig. 11a). In the slurry polymerization the ribbon-like crystals are formed by joined microspheres or platelets, which seem to be laying down side by side in one direction. On the contrary, in the gas-phase polymerization, the platelets turn up in a vertical position stacking themselves one after the other, having the appearance of a shishkebab morphology or fibrillar structure as described by Guttman et al. (5). However, such structures seem to be different in this case due to the following reasons: first, the presence of a core of extended chains could not be detected by DSC techniques, as in the shishkebab case. Second, when the lamellae separated themselves by stretching, microfibers could be observed between them. Under certain conditions the structure was observed having the appearance of a helicoidal twisted ribbon as seen in Fig. 11b. The striation or ribbon thickness ranged between 150-300 Å, corresponding to the repeat distance for the folded chain macroconformation. The chain axes were now oriented lengthwise or parallel to the axes of the worm-like morphology (see also Fig. 13, III-C and III-D).

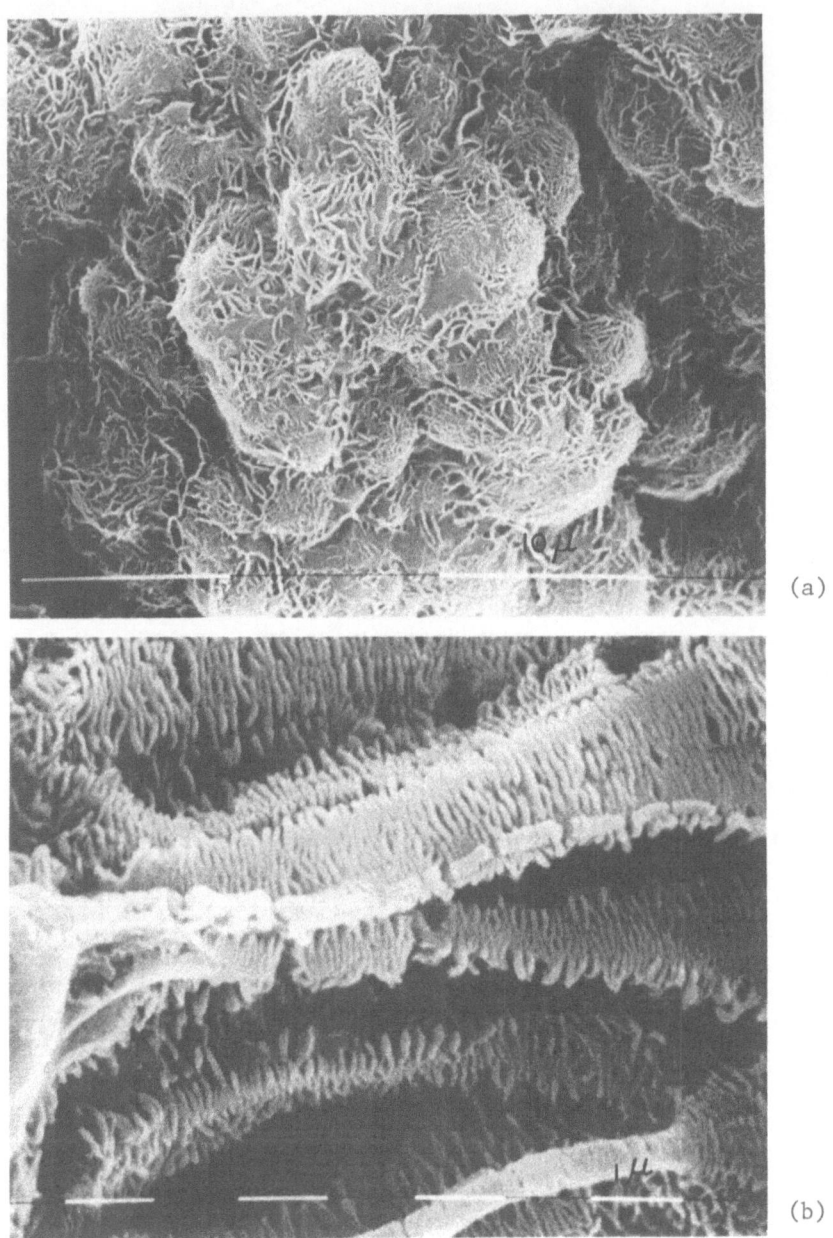

(a)

(b)

Figure 10. Scanning electron micrographs of as-polymerized PE ob-
            tained by gas-phase polymerization at 120°C: a) Nodular
            type morphology, b) worm-like type morphology.

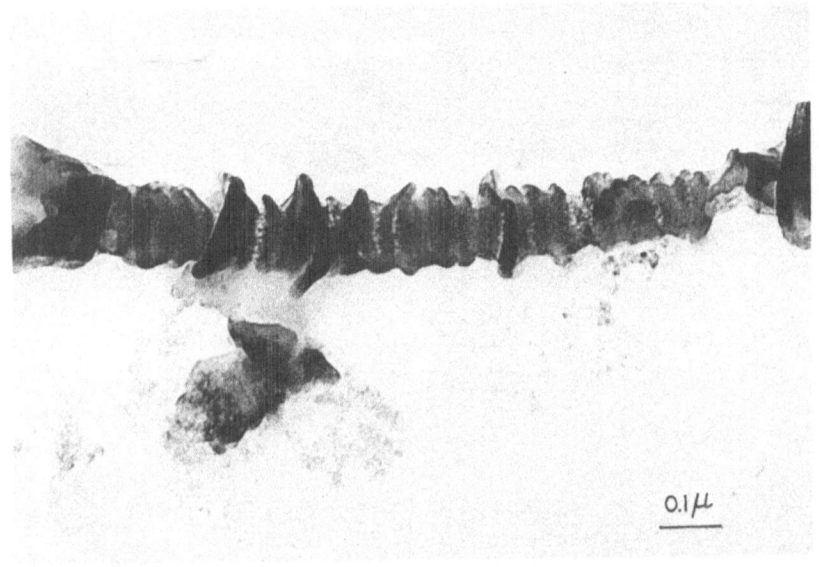

(a)

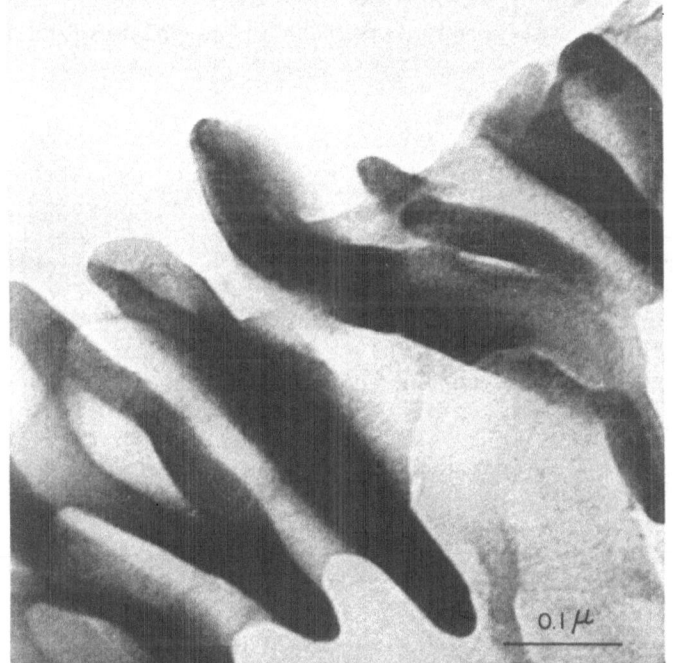

(b)

Figure 11. a) Transmission electron micrographs of as-polymerized
PE obtained by gas-phase polymerization at 120°C; b)
detail of the helicoidal twisted ribbon crystal.

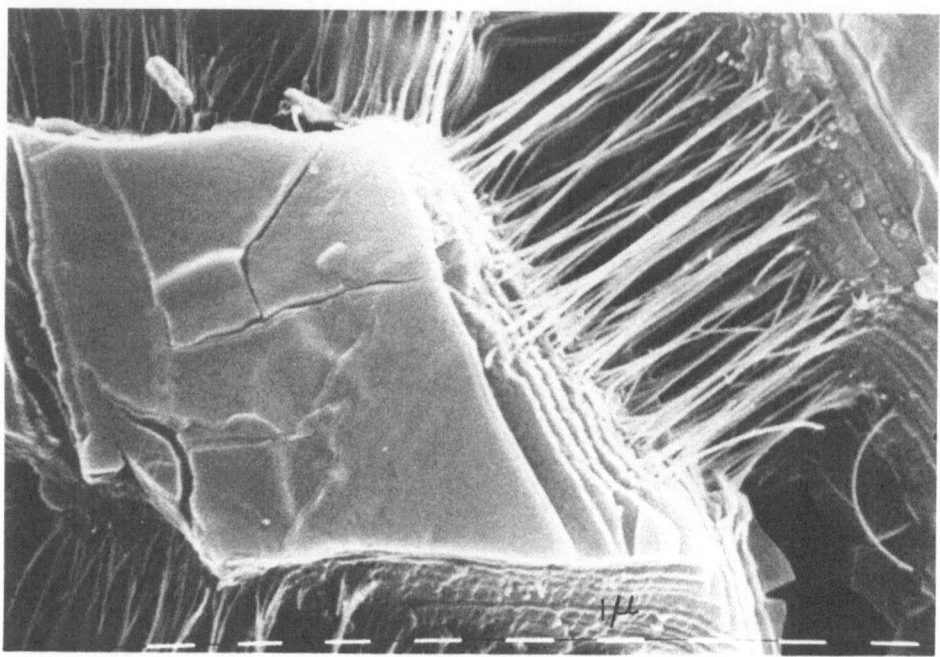

Figure 12.   Scanning electron micrograph of as-polymerized PE obtain-
ed by gas-phase polymerization, using a ratio $|Al|/|Ti|$
of 50:1.

Increasing the ratio $|Al|/|Ti|$ to very high values (50:1), the
main part of the observed morphology was films of polyethylene,
which form several layers covering the catalyst surface. During the
sample preparation for observation under the electron microscope
and by mechanical stress caused by the growing polymer, the catalyst
layers break up, stretching the polymer films into fibrils, as seen
in Fig. 12. The change in the polymer morphology to a rather
structureless film, might be due to the homogenization of the active
sites with different activities, by the drastic reduction of the
catalyst surface and also by the formation of a thin film of polyolefin produced by the excess of alkyl aluminum, as suggested by
Hock (17).

CONCLUSIONS

The following conclusions might be drawn from the above results:
the overall and also the fine morphology of the nascent polyethylenes obtained by the $TiCl_3$ heterogeneous Ziegler-Natta catalyst are
controlled by the catalyst particles rather than by the polymeriza-

STRUCTURE AND MORPHOLOGY OF NASCENT POLYETHYLENES

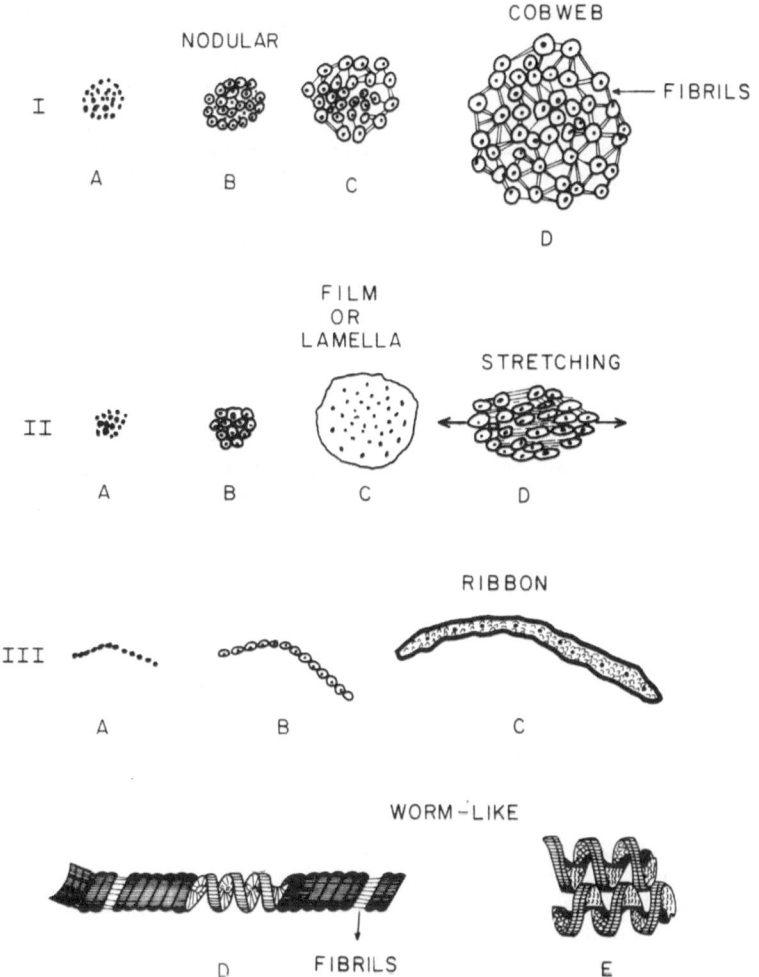

Figure 13.   Stages in the growth of the different morphologies.

tion techniques used to carry out the polymerization.   The catalyst
particles consist of an agglomeration of tiny primary particles.
The resulting polymer morphology depends on the manner in which the
primary catalyst particles are arranged with respect to each other
and also to the migration of the particles during polymerization.
The catalyst particles are probably bonded together by the aid of
the alkyl aluminum.   Therefore, one, two and three dimensional aggregations give rise to the ribbon-like, lamellar and nodular morphologies as represented schematically in Fig. 13 (I, II and III).

The elementary morphology of the nascent polyethylene is basically microspheres or platelets originated by tiny catalyst particles. Thereafter, the different aggregations of this elementary morphology lead to all the types of super-morphologies found. At a low polymerization rate the most common morphology observed is the lamellar, but at a higher polymerization rate the nodular, ribbon-like and worm-like morphologies are predominant. At high polymerization rate the catalyst particles are rapidly covered by the lateral growth of the microspheres or platelets, filling the available surface; the catalyst particles are then most likely lifted and their growth only in the third dimension is now predominant, forming the ribbon-like morphology. The worm-like morphology is formed by the agglomeration of helicoidal twisted ribbon-like crystals giving rise to super-helices (see Fig. 13, III-E).

The cobweb morphology (Fig. 13, I-D) and other fibrillar formations are originated by deformation or stretching of the polymer as the morphologies expand during their growth.

Finally, the nascent morphology of the PE might be to some extent changed by using different ratios of $|Al|/|Ti|$, giving rise to a film morphology at high ratios.

SYNOPSIS

The nascent morphology of the polyethylenes obtained with the heterogeneous $TiCl_3$ Ziegler-Natta catalyst was investigated, using mainly scanning and transmission electron microscopy techniques. Three different polymerization processes were used to carry out the polymerization of the ethylene. Polymerizations in slurry using Et_3Al as co-catalyst and under stirring (10-1000 rpm) were performed in n-paraffine as a solvent, at polymerization temperatures below the solution temperature of the forming polymer (0-70°C) and at pressures between 1-10 atm. Under these conditions the polymer crystallizes during its formation. Four different morphologies were observed: film, globular, ribbon-like and worm-like. The film morphology is formed at lower and higher polymerization rates or temperatures respectively. Polymerizations in solution were carried out at 120°C. At this temperature the nascent morphologies start to dissolve so that a rather smooth layer of polymer was observed. However, as the polymerization proceeds (high polymerization times) a partially dissolved worm-like morphology was also found, indicating that the catalyst produces the polymer at a very high rate and then starts to dissolve. In order to establish whether the presence of a solvent and the application of stirring have any effects on the nascent morphologies or are controlled exclusively by the catalyst, ethylene was also polymerized from the gas-phase (without solvent) and under quiescent conditions (without stirring) using the catalyst system $TiCl_3/Et_2AlCl$. Under these conditions it was possible to observe the nascent morphologies more clearly. The polymer starts forming a film covering the catalyst surface. However, very high active centers give rise to polymer dots which by agglomeration form the ribbon-like crystals. These are looped and helicoidally twisted forming the worm-like morphologies. Finally, a mechanism for the formation of all morphologies is proposed.

REFERENCES

1. B. Wunderlich, "Macromolecular Physics", vols. 1 and 2, Academic Press, New York (1973, 1976).
2. J. Boor, Jr., "Ziegler-Natta Catalysts and Polymerization", Academic Press, New York (1979).
3. H.D. Chanzy, B. Fisa and R.H. Marchessault, Crit. Rev. Macromol. Sci., 1, 315 (1973).
4. L. Rodriguez and J. Gaban, "Macromolecular Chemistry" (J. Polymer Sci., C, 4), M. Magat, ed., Interscience, New York (1964), p. 125.
5. J.Y. Guttman and J.E. Guillet, Macromolecules, 1, 461 (1968); 3, 470 (1970).
6. P. Blais and R. St. John Manley, J. Polymer Sci., 6, 291 (1968).
7. B. Wunderlich, Advan. Polymer Sci., 5, 568 (1968).
8. A. Muñoz-Escalona and A. Parada, Polymer, 20, 474 (1979).
9. Farbwerke Hoechst. British Patent No. 960.232 (1964).
10. L. Lucini, A. Monini, and D. Zaffagnini, in the 37th Annual Technical Conference, Society of Plastics Engineers (SPE), New Orleans (May 1979), p. 506.
11. P. Blais and R. St. John Manley, J. Polymer Sci., 6, 291 (1968).
12. R.H. Marchessault and H.D. Chanzy, J. Polymer Sci., 30, 311 (1970).
13. R.J.L. Graff, G. Kortleve and C.G. Vonk, J. Polymer Sci., 8, 735 (1970).
14. A. Muñoz-Escalona and A. Parada, J. Cryst. Growth, 48, 250 (1980).
15. R.J. Chiang, J. Phys. Chem., 69, 1945 (1965).
16. A. Muñoz-Escalona and A. Parada, Polymer, 20, 859 (1979).
17. Ch.W. Hock, J. Polymer Sci., 4, 3055 (1966).
18. L.A.M. Rodriguez and H.M. van Looy, J. Polymer Sci., 4, 1971 (1966).
19. G. Natta and G. Mazzanti, Tetrahedron, 8, 86 (1960).

CRYSTAL TRANSFORMATION, PIEZOELECTRICITY, AND FERROELECTRIC

POLARIZATION REVERSAL IN POLY(VINYLIDENE FLUORIDE)

Kasumi Matsushige* and Tetuo Takemura

Department of Applied Science, Faculty of Engineering
Kyushu University, Hakozaki, Higashi-ku, Fukuoka, 812
Japan
*Present address: Research Institute for Applied
Mechanics, Kyushu University

SYNOPSIS

    The mechanisms for the II-I crystal transformation in poly(vinyl-
idene fluoride) (PVDF) by various procedures were studied with a PSPC
(position sensitive proportional counter) X-ray system. Simultaneous
X-ray and stress-strain relationship measurements during a drawing
procedure revealed that the crystal transformation from Form II to
Form I always initiates at the deformation stage where a necking was
completed at the center of tensile samples, thus suggesting that a
heterogeneous stress distribution in the sample plays a critically
important role. High pressure X-ray experiments on a heating process
exhibited that this polymer transforms from Form II to folded-chain
Form I and then extended-chain Form I crystals before melting. The
II-I crystal transformation was also observed to proceed with an
activation energy of 30 kcal/mol on an annealing procedure at 4000
kg/cm^2. Furthermore, a uniaxial compressional deformation and a
drawing at high pressures were observed to cause this II-I crystal
transformation. These phenomena were utilized to prepare the Form I
samples with a high degree of crystal perfection and to improve
considerably their piezoelectric properties. Finally, ferroelectric
polarization switching experiments were carried out for Form I crystal
films in wide ranges of temperature and pressure. The switching
current behavior at atmospheric pressure changed remarkably at about
-50°C which coincides well with the reported glass transition
temperature. The pressure dependence of the maximum switching current
time indicted that a polarization reversal mechanism is closely linked
up with the molecular motions in an amorphous region. These observations suggest that there exists a close correlation between the

115

polarization reversal phenomenon in PVDF and molecular chain motions, and further the polarization reversal begins preferentially at particular portions such as a boundary between crystalline and amorphous parts and defects in the sample.

INTRODUCTION

Poly(vinylidene fluoride) (PVDF) is a polymeric material with very interesting scientific and technological properties. This polymer is known to crystallize into three crystal forms, denoted I (β), II (α), and III (γ), depending upon crystallization conditions, and to perform mutual crystal transformations between these crystal forms [1-12]. Several controversial points, however, still remain on the crystal form for the samples melt-crystallized or annealed at high pressures, the crystal structure of Form III, and a role of stress field in the II-I crystal transformation by a drawing procedure. Recently, a fourth crystal form, polar Form II, is reported to be induced from non-polar Form II crystal by an application of a high d.c. electric field [13-17].

Interest in electrical properties of PVDF spouted in 1969 when H. Kawai [18] found that this polymer exhibited an extraordinarily large piezoelectric activity surpassing that of quartz. Soon afterward, Bergman et al. [19] reported on the pyroelectric behavior of PVDF. Because of the combined merits of high elasticity, high processing capability, low acoustic impedence, and high piezoelectricity, PVDF has become one of the most attractive polymeric materials with new functionalities and additional utility values. During the last several years, the number of investigators and their publications on PVDF have shown an enormous increase. And it is now widely recognized that the highest piezoelectric activity is observed for highly oriented Form I crystal samples among several crystal forms in PVDF. The Form I crystal contains a spontaneous polarization in its unit cell and can be usually obtained by the crystal transformation from Form II crystal. Recent X-ray diffraction [14-17] and infrared [13,20,21] experiments have revealed that the orientation of CF_2 dipoles in PVDF is induced by an application of high electric field. Furthermore, the hysteresis in piezoelectric activities during cyclic changes of electric field was reported by several research groups [22-25]. These findings provide a solid proof for a ferroelectric nature of PVDF. Also, theoretical studies were conducted based on various models [26-29], and the piezoelectric constants and piezoelectric relaxations of PVDF observed experimentally were theoretically evaluated with good accuracy.

Despite these active investigations, many unsolved problems still exist even in the basic area, e.g. the detailed mechanism for the II-I crystal transformation, the definite origins of the piezoelectric activity, and the correlation between the ferroelectric polarization reversal phenomenon and molecular chain motions. These unsolved problems have hampered deeper understanding of the unique electric properties of PVDF and development of further applications.

In this study, attention is devoted to solve some of the above mentioned problems and the main emphasis is placed on a prominent role of hydrostatic pressure. A PSPC (position sensitive proportional counter) X-ray system was utilized to observe directly the II–I crystal transformation processes taking place during a drawing procedure at atmospheric pressure, during a heating and an annealing procedure at high pressures, and during a uniaxial compressional deformation. Furthermore, several methods to prepare the oriented Form I crystal films with a high degree of crystalline perfection were demonstrated and the piezoelectric properties of these films were measured. Finally, polarization switching experiments were performed in the wide ranges of temperature and pressure in order to obtain a basic understanding of the polarization switching mechanism. Some of the contents described here have appeared in several publications [30–33].

EXPERIMENTAL

X-Ray Measurements

Figure 1 shows schematically the X-ray measuring system for observing a crystal transformation process taking place during a drawing procedure (a), a heating and an annealing procedure (b), and a uniaxial compressional deformation (c). In order to study this process directly, a strong rotating anode X-ray generator (60 kV, 200 mA) and a PSPC (manufactured by Rigaku Denki Co.) X-ray probe were used. For drawing experiments, a tensile machine was specially designed. A driving force was generated by a synchronous motor, by which the strain rate was kept at a constant value of 5%/min. The sample holders move in opposite directions so that an X-ray beam spot did not transfer from the center of samples even during a drawing process. For high pressure experiments, the sample container was made of Be metal and the details of the apparatus are described elsewhere [34]. The NBS type diamond anvil apparatus (manufactured at High Pressure Diamond Optics Inc.) was modified for the uniaxial compressional experiment. The piston and Be metal anvils were newly constructed for this purpose.

High-Pressure Drawing Apparatus

Figure 2 shows the apparatus used for the drawing at high pressures and high temperatures. A pressurized fluid (silicone oil, 10 cs) was fed to a high pressure vessel through a hole in an upper piston and the value of applied pressure was monitored with a calibrated manganin manometer. Temperature was controlled with a heater

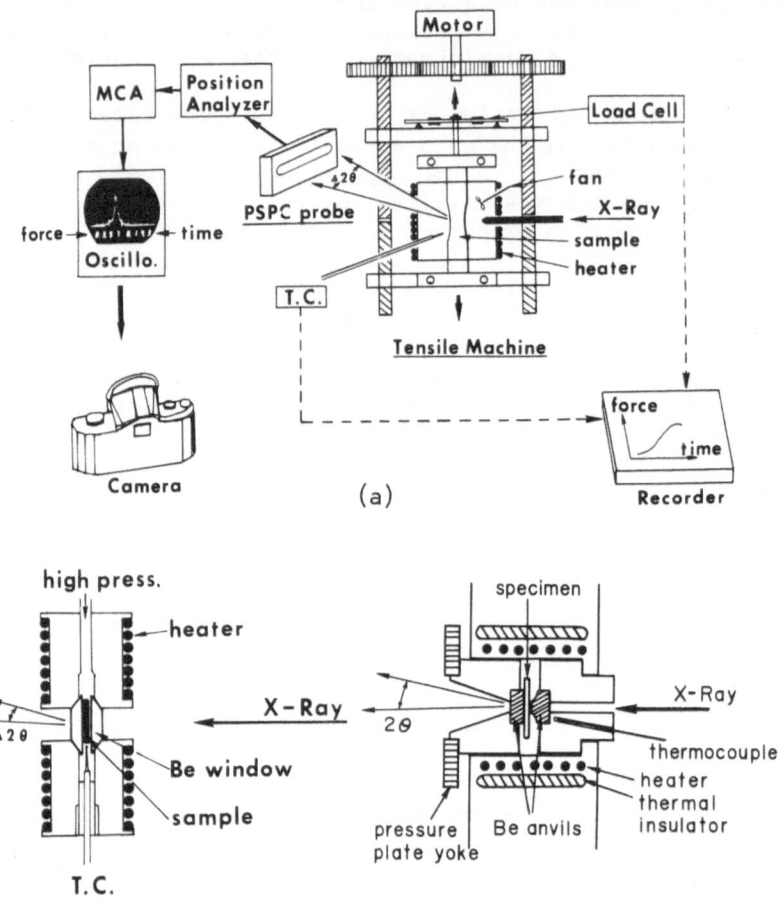

Figure 1:   Schematic diagrams of the X-ray system for observing a
            critical transformation phenomenon during a drawing process
            (a), heating and annealing processes at high pressures (b),
            and a uniaxial compressional process (c).

wound around the pressure vessel and measured with an alumel-chromel
thermocouple. One side of tensile specimen was fixed to the pressure
vessel, and the other to a lower-side piston. Since both upper- and
lower-side pistons were at fixed positions, a tensile force could be
applied to the specimen by shifting the pressure vessel upwards.
Deformation rate was monitored with a differential transformer. In
order to avoid an environmental effect, a wood alloy was employed as a
pressure transmitting fluid especially around the tensile specimen.

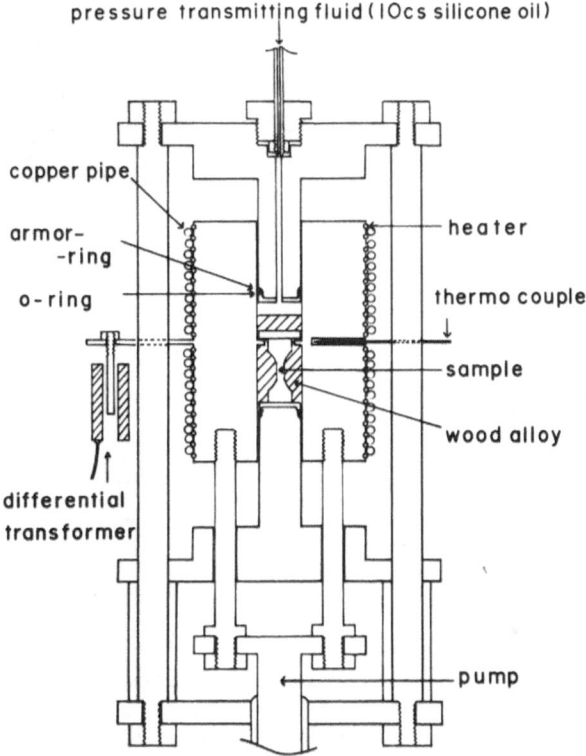

Figure 2. A schematic view of a high pressure and high temperature drawing apparatus.

Piezoelectric Measurements

Measurements of piezoelectric stress constant (e_{31}) and elastic modulus (c) were performed with a Vibron II instrument attached to a charge amplifier. A piezoelectric strain constant (d_{31}) was calculated from the relation d = e/c. A poling procedure was carried out at 20 MV/m and at 125°C for 30 min. Aluminum was evaporated on both surfaces of samples as electrodes before the poling procedure.

Polarization Switching Current Measurements

Polarization switching current measurements were performed according to the technique first employed by Merz [35]. A schematic diagram of the experimental system is shown in Figure 3. High d.c. voltage steps were applied to film samples through a switching device

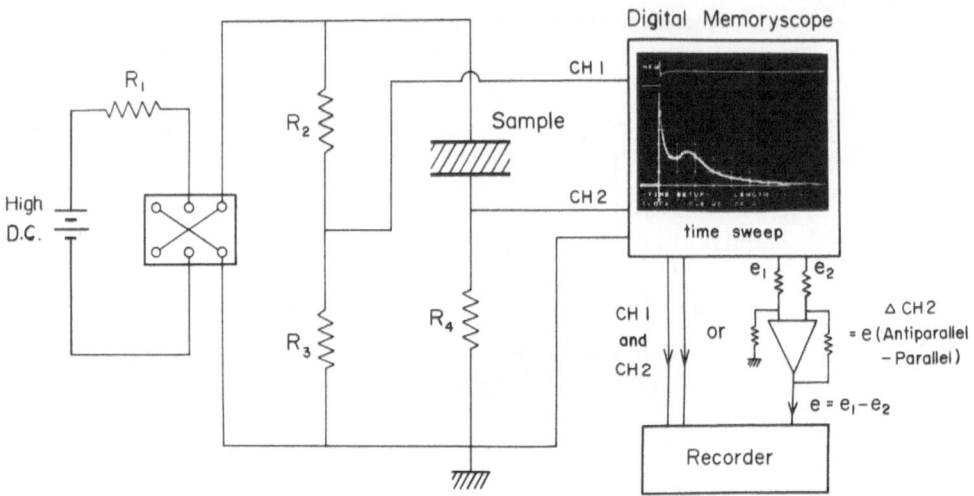

Figure 3:   A schematic diagram of polarization switching experimental system.

consisting of 18 pieces of reed relays connected in series, and then a
polarization switching current was observed as the voltage change
appearing at the ends of resistance R_4 with a digital memory scope
(Iwatsu, DMS-6430). Film samples were sandwiched between a pair of
brass anvils, which were supported with a spring to obtain close
contact with the samples. The measurements both at atmospheric
pressure and at high pressures were carried out by setting a sample
cell into a high pressure vessel. Silicone oil was used as a pressure
transmitting fluid and also served for preventing a discharge phenomenon from taking place between electrodes. The details of the
experimental procedure are explained elsewhere [33].

Compressibility Measurements

        Compressibility measurements were performed in two ways. A high
pressure X-ray diffraction apparatus with a PSPC probe (Figure 1a) was
employed to obtain the compressibility of a crystalline part in the
PVDF sample. The compressibility of the whole sample (bulk compressibility) was measured with a high pressure dilatometer [36]. Then,
the compressibility of an amorphous part was calculated by subtracting
the contribution of the crystalline part to the bulk compressibility
with the consideration of the crystallinity of the used sample.

Sample Preparation

        The original sample used in this study was KF-polymer supplied by
Kureha Chemical Industry Co., Ltd. The number-average molecular

weight was 54,000 and the regular (head-to-tail linkage) content was 90.3% estimated by high resolution ^{19}F nuclear magnet resonance (data of Kureha Chemical Industry Co., Ltd.). For drawing and uniaxial compression experiments, KF-polymer films (50 μm and 500 μm in thickness) were used as original Form II samples. These films were confirmed by X-ray and infrared measurements to be Form II crystal sample with almost no Form I crystal content. For polarization reversal experiments, the film samples were uniaxially drawn up to five times its original length at 75°C and at atmospheric pressure and then heat-treated with its ends fixed at 130°C for one hour.

RESULTS AND DISCUSSION

Crystal Transformation During Drawing at Atmospheric Pressure

Simultaneous drawing and X-ray diffraction measurements were performed at atmospheric pressure in the temperature range of 75°C - 150°C. Figure 4 shows a nominal stress-strain curve observed at 75°C and the X-ray diffraction patterns taken at various deformation stages. The observed stress-strain relationship revealed that the sample underwent yielding, necking, and strain hardening as its strain increased. Also, the observed changes in the X-ray diffraction patterns revealed the detailed crystallographic nature of the crystal transformation from Form II to Form I, as described below. An appearance of Form I (200) and (110) combined reflection was detected at stage 3, as indicated by an arrow in the third picture. At this deformation stage the necking was initiated at the center of a tensile sample and started to propagate to both sample ends. With further increasing strain, an apparent peak angle of the main reflection exhibited a gradual shifting from 20.1° (the diffraction angle of Form II (110) reflection), to 20.8° (the one expected for Form I (200) and (110) combined reflection), and simultaneously a narrowing of its half-height width. In contrast to the progressive development and completion of Form I crystal during the drawing process, the other (020), (100), and (021) reflections from Form II crystal diminished in their intensities with increasing strain.

Similar drawing experiments were performed at different temperatures between 75°C and 150° C. The results are summarized in Figure 5, where the nominal stress-strain curves observed at various temperatures and the information detected from X-ray direction experiments are shown. The crystal transformation from Form II to Form I was observed in the low temperature region below 130°C but not in the high temperature region above 140°C. It was also revealed that the degree of crystal conversion from Form II to Form I depended on the drawing temperature and decreased with increasing drawing temperature. As indicated by arrows in the figure, it is characteristic that the crystal transformation always initiated at the deformation stage where the necking was initiated at the center of the tensile sample.

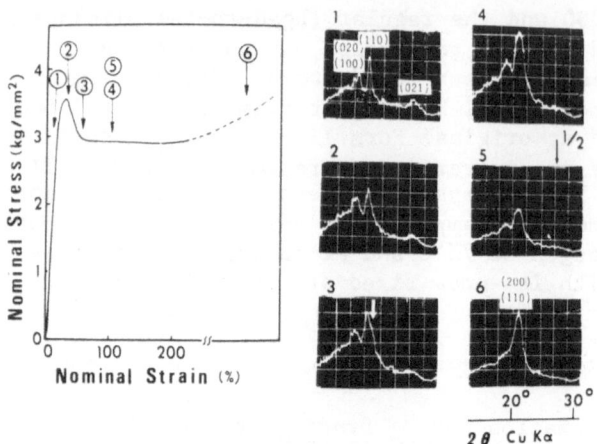

Figure 4:   A nominal stress-strain curve and X-ray diffraction
            patterns observed during a drawing experiment at 75°C. The
            X-ray diffraction pictures 1-6 were taken at the deforma-
            tion stages indicated by number 1-6 on the nominal stress-
            strain curve, respectively.

        As discussed above, a criterion determining whether the crystal
transformation occurs or not seems to exist especially in the observed
nominal stress-strain curves. That is, when the sample performs so-
called "cold drawing", the crystal transformation takes place, whereas
when the sample exhibits "hot-drawing", it does not. Therefore, it
has been observed that the drawing mechanism bears close correlation
to the crystal transformation phenomenon and that the heterogeneous

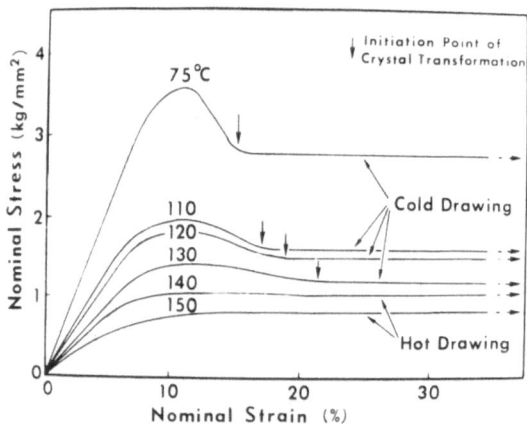

Figure 5:   The nominal stress-strain curves observed at different
            temperatures between 75°C and 150°C.

deformation process which occurs in the cold drawing plays a critically important role in the crystal transformation phenomenon.

Crystal Transformation During Heating and Annealing at High Pressure

     At high pressures, DTA curves for Form II crystal exhibit double
or triple endothermic peaks and an exothermic peak on heating [9].
This fact suggests that PVDF forms new crystals in the melting region
different from the original Form II crystal. Figure 6 shows X-ray
diffraction patterns displayed on an oscillograph and the DTA curve
observed at 4000 kg/cm^2. Both these experiments were performed at the
same heating rate of 5°C/min to obtain a correct correspondence
between the DTA curve and the structural change indicated in the X-ray
diffraction patterns. The X-ray reflections characteristic of Form II
crystal are clearly seen at 250°C, but their intensities become weaker
at 278°C. At 289°C, which is slightly higher than the peak temperature
ture of the lowest DTA endothermic peak, X-ray reflections from Form
II crystal disappear. On the other hand, a new X-ray reflection at
$2\theta \approx 21°$ suddenly appears when the temperature is further raised to
297°C, which corresponds to a temperature higher than the exothermic
peak temperature. The reflection angle of $2^\theta \approx 21°$ is expected for

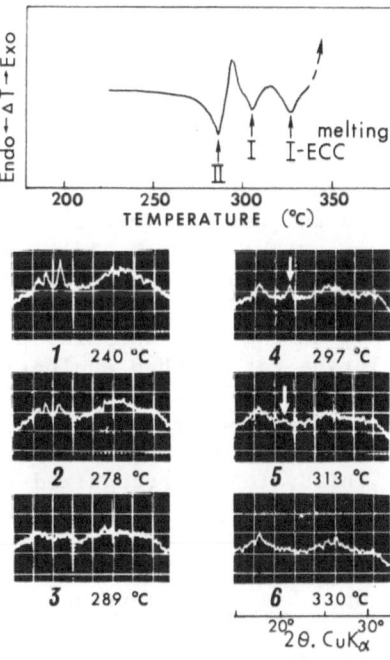

Figure 6: An upper part of figures shows a DTA thermogram observed
           at 4000 kg/cm^2. A lower part exhibits the X-ray diffrac-
           tion patterns observed at 4000 kg/cm^2 and at the tempera-
           tures corresponding to different melting stages.

(110) and (200) combined reflection of Form I crystal. Only a trace of this reflection can be detected at 313°C. Finally, the X-ray diffraction pattern at 330°C exhibits an amorphous peak centered around $2\theta = 17.5°$, reflecting the molten state of PVDF sample. The observed changes in X-ray diffraction patterns with increasing temperature provide a reasonable explanation of the multiple DTA peaks at 4000 kg/cm^2. That is, an endothermic peak at 286°C is due to melting of the Form II crystal used as a starting sample. The exothermic peak at 294°C is due to the crystal formation of Form I. On the other hand, endothermic peaks at 305°C and 326°C are thought to be due to melting of Form I folded-chain crystal and Form I extended-chain crystal, respectively. The formation of Form I extended-chain crystal was probed indirectly by the examination for the sample melt-crystallized or annealed at high pressures with optical and electron microscopies [9].

Next, the time effect on the crystal transformation under high pressure is discussed. The preceding section dealt with the crystal transformation phenomenon taking place with a rather fast heating rate of 5°C/min. If the measuring time scale is changed, the crystal transformation process alters its nature. Since at high pressures Form I crystal is more stable than Form II crystal, it is probable to expect that the crystal transformation from Form II to Form I will occur at a temperature even lower than the melting point of Form II crystal. In order to ascertain this, changes in X-ray diffraction pattern were followed with time under constant conditions of pressure and temperature. Annealing experiments were performed at 4000 kg/cm^2 and Figure 7 shows the time dependence of the amount of Form I crystal (N_1) at various temperatures estimated from the observed X-ray diffraction patterns. The time dependence of N_1 depends drastically on annealing temperature and the rate of increase in N_1 is enhanced with increasing annealing temperature. It should be noticed that the annealing temperatures employed here are lower than the melting point of Form II crystal at 4000 kg/cm^2 (286°C). Thus, this II-I crystal transformation phenomenon during the annealing procedure is thought to be of diffusion control-type rather than nucleation control-type. Arrhenius plots of the relaxation time for the II-I crystal transformation which is as the simplest approximation calculated from initial slopes in the log N_1 vs. time curves shown in Figure 7 gave the value of about 30 kcal/mol. This value is recognized to be an activation energy for PVDF to undergo the crystal transformation from Form II structure to Form I structure at the pressure of 4000 kg/cm^2.

Crystal Transformation Upon Uniaxial Compression

The direct observations of structural changes occurring during a uniaxial compressional deformation were performed at room temperature using a Be anvil compression apparatus (Figure c). Figure 8 shows X-ray diffraction patterns under different compressional stress conditions. At atmospheric pressure X-ray reflections from Form II

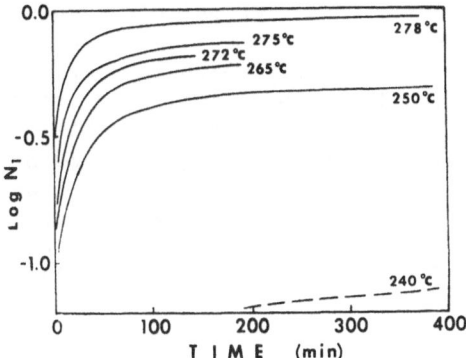

Figure 7:   Time dependences of N_1, the content of Form I crystal, during the annealing process at 4000 kg/cm^2 at various temperatures.

crystal as a starting sample was clearly detected. When the compressional stress was increased step by step and the X-ray diffraction pattern was examined at each occasion, a detectable decrease in intensities of Form II (100) and (110) reflections was observed at 3000 kg/cm^2. With further increase in the applied compressional

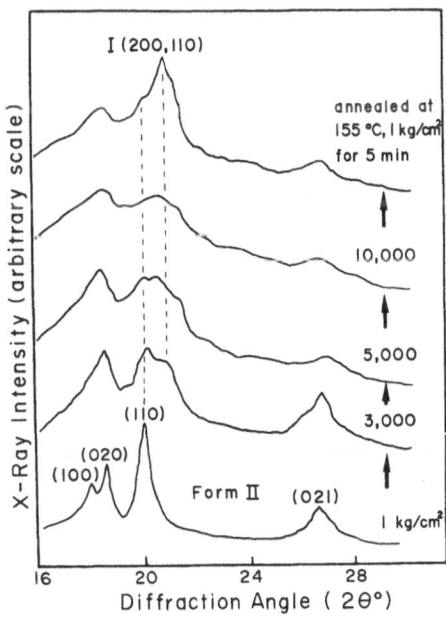

Figure 8:   The X-ray diffraction patterns observed at various compressional stress conditions.

stress, these reflections almost disappeared and other Form II (020) and (021) reflections became broad and less intense. On the other hand, a new reflection at $2\theta \simeq 21°$, which corresponds to Form I (200) and (110) combined reflection, could be clearly observed when the sample was annealed for 5 min at the condition of 155°C at atmospheric pressure. The X-ray diffraction patterns observed at high compressional stresses (5000 and 10,000 kg/cm^2) were diffuse due to the situation that the sample compressed with the Be anvils (the shape of which was circular) was deformed in radial directions. However, additional infrared and X-ray fiber pattern examinations for the samples compressed with circular or rectangular shaped anvils confirmed that the uniaxial compressional deformation cause the II-I crystal transformation. When the rectangular shaped anvils were used, the X-ray fiber pattern from the resultant sample was very clear, with a degree of c-axis orientation of about 95%, suggesting that the sample deformed almost uniaxially perpendicular to the major direction in the rectangle due to directional friction between the anvils and film surfaces, as demonstrated in Figure 9.

The details of the II-I crystal transformation mechanism in the compressional deformation was not studied since a compressional stress-strain relationship measurement is very difficult to carry out simultaneously with the X-ray diffraction measurements. However, the facts that the compressional deformation is of shear-type and the deformation at high temperatures above 150°C resulted in a lower degree of the II-I crystal conversion ratio suggest to us a similar mechanism as the case of the drawing is valid also in this case. This compressional deformation method has various advantages. The procedure is very simple and it is easy to produce the film samples with

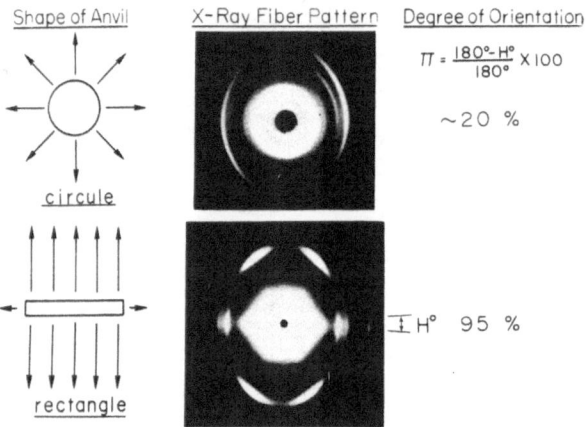

Figure 9: The influence of the shape of compressional anvils on deformation modes of samples. The samples were compressed at the same condition of 5000 kg/cm^2 and 125°C.

desired shapes if suitable shapes of compressional anvils are employed. Furthermore, by insulating the anvils from each other electrically, high d.c. voltage can be applied to the samples even during a compressional cycle. This simultaneous poling procedure during the II-I crystal transformation produced the film samples with a very high piezoelectricity, as described later.

Crystal Transformation by High Pressure Drawing

With the attempt to get more perfect oriented Form I crystal samples, high pressure drawing experiments were performed at 4000 kg/cm^2, at which the II-I crystal transformation phenomenon during an annealing procedure was fully analyzed as described above. The strain rate and final deformation ratio were about 25%/min and five times its original length, respectively. The obtained samples were examined at atmospheric pressure and Figure 10 shows X-ray fiber patterns and X-ray diffraction profiles on an equatorial line located around 2θ (Cu K$_\alpha$) = 20°, and IR spectra in the range of 600 - 1000 cm^{-1} for the samples drawn at different temperatures of 165, 200, 250, and 265°C and at the constant pressure of 4000 kg/cm^2, and the sample drawn at 75°C and at atmospheric pressure, which was employed for a comparison. The fact that the drawing at atmospheric pressure and high temperatures above 130°C scarcely cause the II-I crystal transformation by drawing may not be achieved at temperatures higher than 230°C. However, the experimental results revealed that the high pressure drawing even at 250 and 265°C produced an oriented Form I crystal, probably due to a superimposed annealing effect at high pressures for the II-I crystal transformation with the drawing process. With increasing drawing temperature, the degree of c-axis orientation decreased monotonously from 92% to 88%, but a residual crystal content of Form II estimated from IR 766 cm^{-1} band which is assigned to bending mode of CF$_2$ and CCC in Form II crystal [8] was observed to increase. On the other hand, a half-height width of Form I crystal (200) and (110) combined reflection exhibited the minimum value for the sample drawn at 250°C. Compared with these samples drawn at 4000 kg/cm^2, the one drawn at 75°C and atmospheric pressure exhibited a high degree of c-axis orientation but a larger half-height width value and much more residual content of Form II crystal.

Piezoelectric Properties

The II-I crystal transformation phenomena described above were utilized to improve the piezoelectric properties of PVDF. Table I compares the piezoelectric constants for the samples prepared by various procedures. If we employ e$_{31}$ value of the sample drawn at 75°C and at atmospheric pressure as a standard, the samples prepared by an annealing and a drawing at high pressures, and a uniaxial compression give about 3.5, 4.1 (the maximum), and 2.8 times higher values, respectively. This is thought to be due to various reasons such as improvements in crystallinity and crystal perfection of Form I

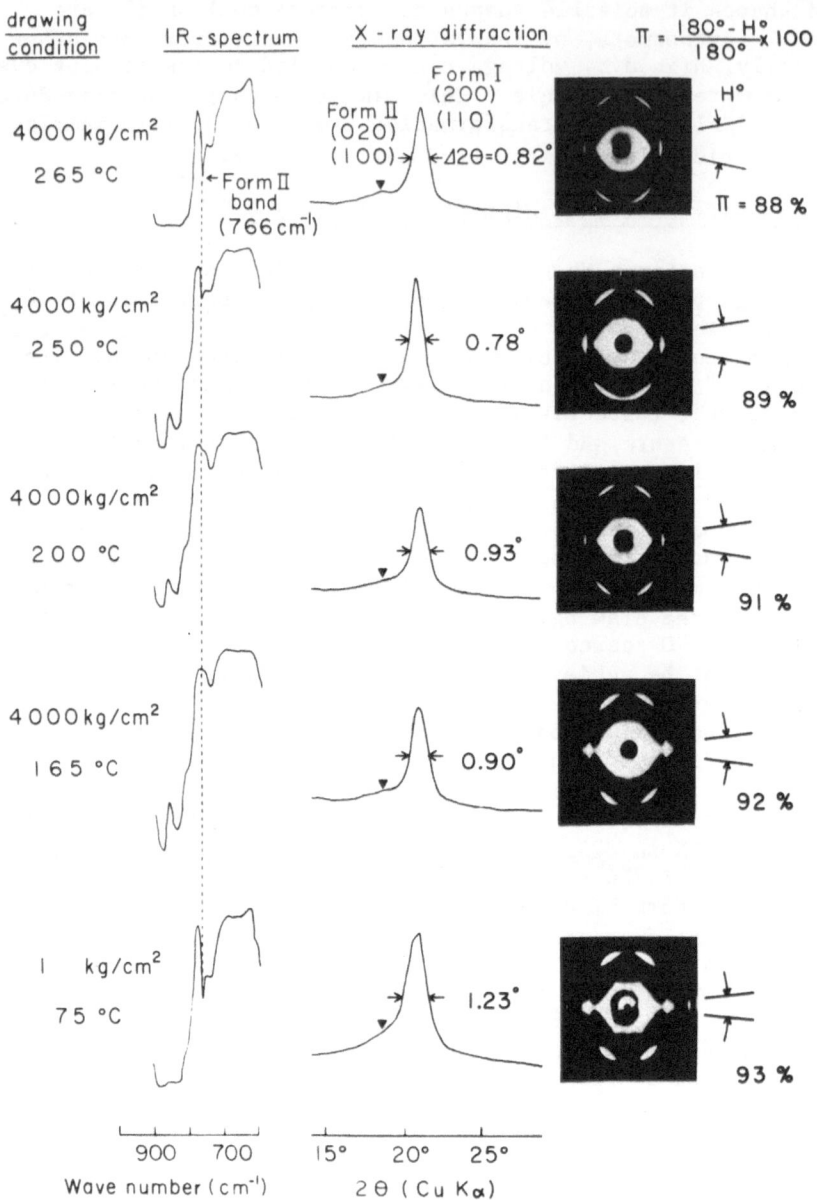

Figure 10: X-ray (equatorial and fiber) diffraction patterns and infrared spectra observed for the samples drawn at various conditions.

Table I

Comparison of piezoelectric properties of the samples prepared by various methods. A poling was performed at 20 MV/m and at 125°C for 30 min.

sample preparations		piezoelectric stress const. e_{31} (10^{-2} C/m^2)	elastic modulus c (10^9 N/m^2)	piezoelectric strain const. d_{31} (pC/N)
drawing at 1 kg/cm^2 and 75°C		1.0	1.5	6.7
high pressure annealing (4000 kg/cm^2, 278°C, and 30 min)		3.5	1.3	27
high pressure drawing (at 4000 kg/cm^2)	at 165°C	2.7	1.4	20
	at 200°C	3.0	1.4	21
	at 250°C	4.1	1.6	26
	at 265°C	1.7	1.3	13
uniaxial compression	poling after compression	1.2	(1.5)	(8.0)
(at 5000 kg/cm^2 and 125°C)	simultaneous poling	2.8	(1.5)	(19)

crystal. Some information on the structural changes induced by the high pressure treatments were obtained by X-ray diffraction and infrared examinations. Compared with the sample drawn at 75°C and at atmospheric pressure, the samples prepared by the high pressure techniques exhibited sharper Form I (200) and (110) combined reflection, reflecting the higher degree of crystal perfection. IR examination, on the other hand, revealed that the thermal decomposition taking place during the drawing and annealing procedures at high pressures occurs rather selectively at irregular portions in the samples, as judged from a considerable reduction in the intensities of 1675, 1329, and 1451 cm^{-1} bands of hetero-linkages and of 489 cm^{-1} band of a gauche conformation involved in a non-crystalline phase [8]. These changes are considered to contribute to better alignment of molecular dipoles upon an application of d.c. electric field, and thus to the observed improvement in the piezoelectricity. However, further

detailed studies are necessary for a full understanding of the correlation between the piezoelectricity and the crystal structures.

The one other thing to be noticed is the fact that a simultaneous
poling procedure during a compressional deformation brings about 2.3
times higher e_{31} value than the poling procedure carried out separately after the deformation. This significant improvement is probably
due to the situation that an applied electric field acts very effectively on PVDF molecules which are in a rather unstable state of the
II-I crystal transformation process and so a high degree of dipole
alignment may be achieved. A similar phenomenon that the simultaneous
II-I crystal transformation and poling procedure improves the piezoelectric properties greatly is reported by Fukada et al. [37] for the
case of the drawing under a corona discharge.

Ferroelectric Polarization Reversal

Temperature effects on the ferroelectric polarization switching
behavior of Form I crystal were investigated at atmospheric pressure
over the temperature range from -76°C to 125°C. At room temperature,
the polarization switching behavior depended on an amplitude of an
applied electric field E and notable polarization switching currents,
the peaks of which (τ_{max}) located around a few ms, were detected above
a critical value of $E_c \cong$ 75 MV/m. These two characteristic values of
τ_{max} and E_c observed for PVDF are both about one thousandfold the
values reported for common ferroelectric crystals such as $BaTiO_3$ [35]
and Rochelle salts [38]. This marked difference may be attributed to
the fact that the polarization reversal in PVDF must involve the rotation of molecular chains of larger crystalline units and/or conformational changes of molecular chains requiring a large driving energy,
while that in common ferroelectric crystals can be achieved easily
either by movement of hydrogen atoms or by displacement of ionic atoms
in crystalline latices. Figure 11 shows the time dependences of the
switching current density (j-t curves) observed at different temperatures and at a constant electric field E = 200 MV/m. At -60°C, the
j-t curve is very broad and exhibits its peak around 2.7 ms. With
increasing temperature, the peak position shifts toward shorter times
such as 1.1 ms at -20°C, and 0.4 ms at 0°C, and 0.2 ms at 110°C.

Figure 12 shows the temperature dependencies of the maximum
switching current time (τ_{max}), the maximum switching current density
(j_{max}) which both are defined at the peak position in the observed j-t
curves, and the polarization change Q ($= \int_0^\infty j(t)dt$). The Q values
steeply increase around -60°C, then become almost constant, and
finally decrease gradually above 60°C. This temperature dependence of
the Q value is quite similar to that of the piezoelectric constant
reported by Tamura et al. [39] and Furukawa et al. [27], suggesting
that the piezoelectricity in PVDF has a close correlation with a
remnant polarization. On the other hand, the j_{max} values show a

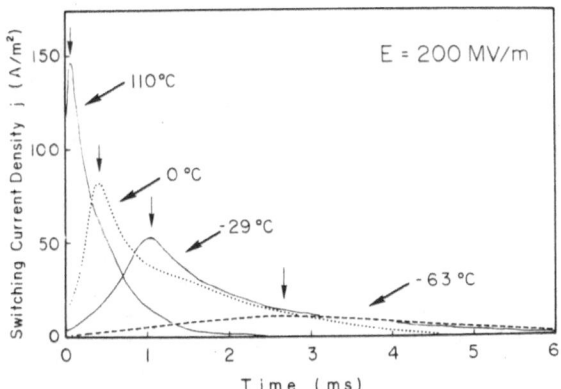

Figure 11: The switching current profiles observed at E = 200 MV/m
and at different temperatures.

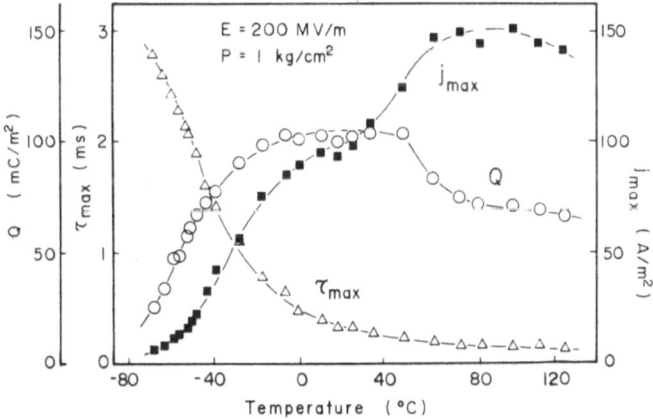

Figure 12. The temperature dependences of τ_{max} (a maximum switching
time), j_{max} (a maximum switching current density), and
Q (a polarization change).

temperature dependence similar to that of the Q values except in the
temperature range from 40°C to 60°C. A substantial decrease in the
τ_{max} values is seen around –50°C. It is notable that the temperatures
of about –50°C and 60°C where the changes in the polarization reversal
characteristics were observed coincide well with the reported β (glass
transition) and α (crystalline) relaxation temperatures [40–43],
respectively, thus revealing again that the polarization switching
phenomenon in PVDF has a close correlation with the molecular motions.

Pressure effects on the polarization switching behavior were
studied at room temperature in the pressure range up to 5000 kg/cm^2.
As shown in Figure 13, the j–t curves observed at E=200 MV/m change
greatly with applied pressures. With increasing pressure, the peak
positions in the j–t curves shift toward longer times and the switching current profiles become broad, revealing that the response of CF$_2$
dipoles to a sudden application of an electric field becomes slower at
high pressures. Figure 14 shows the pressure dependences of three
characteristic values of τ_{max}, j_{max}, and Q. The Q values exhibit an
almost constant value up to about 2500 kg/cm^2 and then a negative
pressure dependence above this pressure. On the other hand, both the
j_{max} and τ_{max} values exhibit the changes in the pressure dependences
around the same pressure of 2500 kg/cm^2.

An application of hydrostatic pressure causes a volume contraction with different degrees to crystalline and amorphous parts in the
test sample. This fact may be utilized to distinguish which part

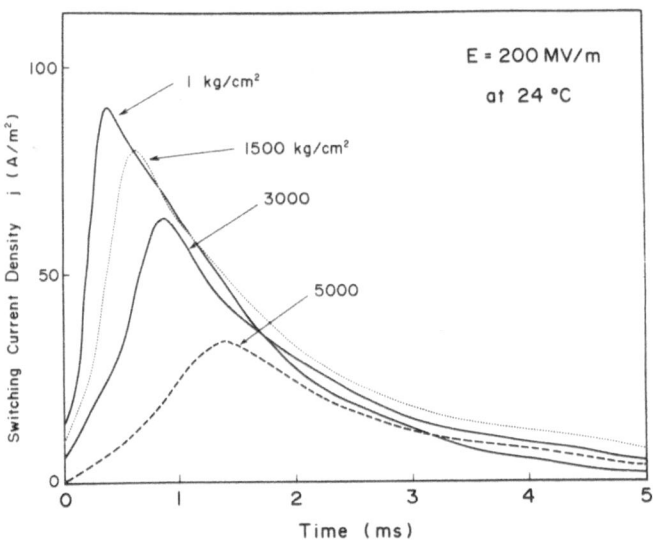

Figure 13:   The switching current profiles observed at E = 200 MV/m
             and at different pressures.

contributes primarily to the observed pressure dependence of the polarization switching behavior. In order to examine this, compressibility measurements for the crystalline and amorphous parts were performed. Figure 15 shows the volumes of bulk, crystalline part, and amorphous part at high pressures, V(P), the values of which are normalized with those at atmospheric pressure, V_o. Among these three cases, the amorphous part exhibits the highest compressibility.

In order to investigate the correlation between the polarization switching behavior at high pressures and the observed volume changes, $V(P)/V_o$, the following equations are derived. If the polarization switching time is assumed to be determined by the Doolittle-type equation [44],

$$\tau \propto \exp(-1/f) \tag{1}$$

where f is a free volume fraction, and the f values change with pressure following the next equation

$$f(P) = f_o \cdot V(P)/V_o \tag{2}$$

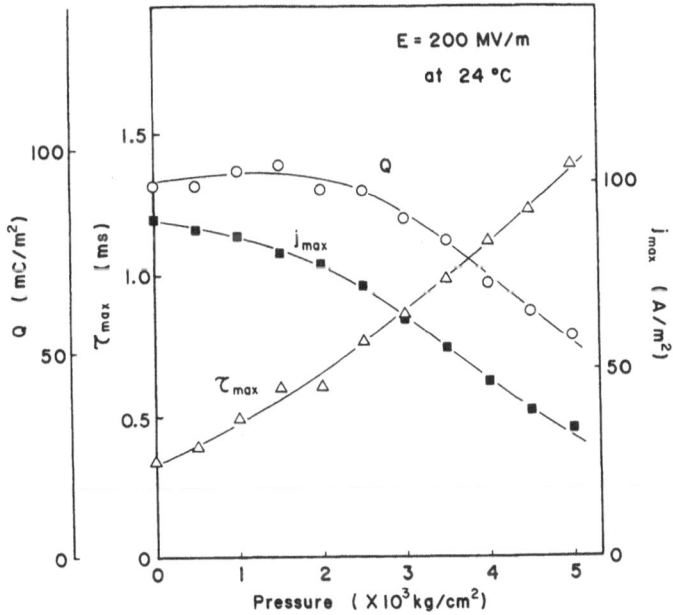

Figure 14:   The pressure dependences of τ_{max}, j_{max}, and Q. The data are taken from the switching current profiles shown in Figure 13.

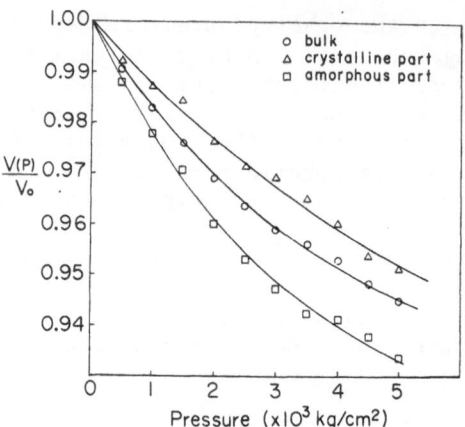

Figure 15:   The pressure dependences of the volumes of bulk, crystal-
             line part, and amorphous part. The values are normalized
             with those at atmospheric pressure.

where f_0 is the f value at the condition of atmospheric pressure and
room temperature. By combining the equations (1) and (2), the τ value
at pressure P, $\tau(P)$, can be expressed by equation (3).

$$\ln \tau(P) = A + V_0/f_0 \cdot V(P) \tag{3}$$

In this study, the τ_{max} values shown in Figure 14 are employed as the
value, and $\ln \tau_{max}(P)$ vs. $V_0/V(P)$ plots are shown in Figure 16.   The

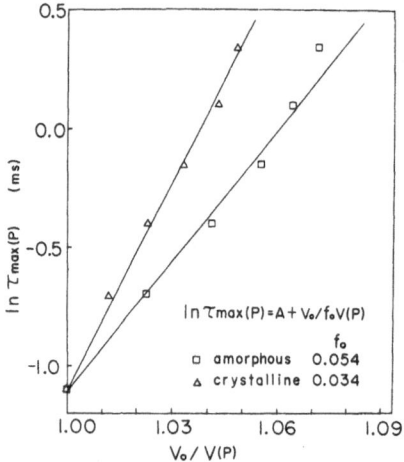

Figure 16:   The relationship between $\ln \tau_{max}$ (P) and $V_0/V(P)$ for
             crystalline and amorphous parts.

data for both the crystalline and amorphous parts locate almost on straight lines and from their slopes f_0 values are calculated to be 0.034 for the crystalline part 0.054 for the amorphous part, respectively. Although the f_0 value of PVDF is not experimentally obtained, this is supposed to be in the range of 0.05 - 0.07 if PVDF contains a similar amount of the free volume fraction as common amorphous polymer [45]. Although it might be necessary to ascertain the validity of the above assumptions, the facts that the pressure dependence of τ_{max} can be well explained by the equation (3) and furthermore its slope gives a reasonable f_0 value for the case of amorphous part suggest to us that the polarization reversal mechanism in PVDF is closely linked up with the molecular motions of the amorphous part rather than the crystalline part, that is, the polarization reversal begins preferentially at the portions such as a boundary between the crystalline and amorphous parts and/or defects in the sample. This view is also supported by the observations about the effect of temperature, as described above, and the influences of heat-treatments and ^{60}Co radiation [46] on the polarization reversal phenomenon.

ACKNOWLEDGMENTS

This work was supported in part by a Grant-in-Aid for Science Research from the Ministry of Education. We are grateful to Messrs. K. Nagata, K. Tagashira, and S. Imada for their assistance with the experiments.

REFERENCES

1. Yu. D. Kondrashov, Gipkh'a, 46, 166 (1960).
2. Ye. L. Gal'Perin, Yu. V. Strogalin, and M.P. Mlenik, Vysokomol. Soedin., 7, 933 (1965).
3. N. I. Makarevich and V. N. Nikitin, Vysokomol. Soedin., 7, 1673 (1965).
4. J. B. Lando, H. G. Olf, and A. Peterlin, J. Polym. Sci., A-1, 4, 941 (1966).
5. W. W. Doll and J. B. Lando, J. Macromol. Sci., Phys., B2, 219 (1968); ibid. B4, 889 (1970).
6. R. Hasegawa, M. Kobayashi, and H. Tadokoro, Polymer J., 3, 591 (1972).
7. R. Hasegawa, Y. Takahashi, Y. Chatani, and H. Tadokoro, Polymer J., 3, 300 (1972).
8. M. Kobayashi, K. Tashiro, and H. Tadokoro, Macromolecules, 8, 158 (1975).
9. K. Matsushige and T. Takemura, J. Polym. Sci., Phys. Ed., 16, 921 (1978).
10. S. Weinhold, M. H. Litt, and J. B. Lando, J. Polym. Sci., Polym. Lett. Ed., 17, 585 (1979).
11. A. J. Lovinger and H. D. Keith, Macromolecules, 12, 919 (1979).

12. Y. Takahasi and H. Tadokoro, Macromolecules, 13, 1317 (1980).
13. P. D. Southgate, Appl. Phys. Lett., 28, 250 (1976).
14. G. T. Davis, J. E. McKinney, M. G. Broadhurst, and S. C. Roth, J. Appl. Phys., 49, 4998 (1978).
15. D. K. Das-Gupta and K. Doughty, J. Appl. Phys., 49, 4601 (1978).
16. R. C. Kepler and R. A. Anderson, J. Appl. Phys., 49, 1232 (1978).
17. B. A. Newman, C. H. Yoon, K. D. Pae, and J. I. Scheinbeim, J. Appl. Phys., 50, 6095 (1978).
18. H. Kawai, Jpn. J. Appl. Phys., 8, 975 (1969).
19. J. G. Bergman, G. R. Crane, A. A. Ballma, and H. M. O°Bryant, Jr., Appl. Phys. Lett., 21, 497 (1972).
20. M. Date and E. Fukada, Rep. Prog. Polym. Phys. Jpn., 20, 339 (1977).
21. D. Naegele and D. Y. Yoon, Appl. Phys. Lett., 33, 132 (1978).
22. M. Tamura, K. Ogasawara, N. Ono, and S. Hagiwara, J. Appl. Phys., 45, 3768 (1974).
23. M. Oshiki and E. Fukada, J. Mater. Sci., 10, 1 (1975).
24. K. Ogasawara, K. Shiratori, and M. Tamura, Rep. Prog. Polym. Phys. Jpn., 19 (1976).
25. J. I. Sheinbeim, C. H. Yoon, K. D. Pae, and B. A. Newman, J. Appl. Phys., 51, 5156 (1980).
26. M. G. Broadhurst, G. T. Davis, J. E. McKinney, and R. E. Collins, J. Appl. Phys., 49, 4992 (1972).
27. T. Furukawa, J. Aiba, and E. Fukada, J. Appl. Phys., 50, 3615 (1979).
28. R. Harakawa and Y. Wada, Adv. Polym. Sci., 11, 1 (1973).
29. K. Tashiro, M. Kobayashi, H. Tadokoro, and E. Fukada, Macromolecules, 13, 691 (1980).
30. K. Matsushige, K. Nagata, and T. Takemura, Jpn. J. Appl. Phys., 17, 467 (1978).
31. K. Matsushige, K. Nagata, S. Imada, and T. Takemura, Polymer, 21, 1391 (1980).
32. K. Matsushige and T. Takemura, J. Cryst. Growth, 48, 343 (1980).
33. K. Matsushige, S. Imada, and T. Takemura, Polymer J., 13, in press.
34. M. Yasumiwa, R. Enoshita, and T. Takemura, Jpn. J. Appl. Phys., 15, 1421 (1976).
35. W. J. Merz, J. Appl. Phys., 27, 938 (1956).
36. T. Ide, S. Taki, and T. Takemura, Jpn. J. Appl. Phys., 16, 647 (1977).
37. T. Goho, T. Furukawa, M. Date, T. Takamatsu, and E. Fukada, Polymer Preprints, Japan, 28, 447 (1979) (in Japanese).
38. H. H. Wieder, Phys. Rev., 110, 29 (1958).
39. M. Tamura, S. Hagiwara, S. Matsumoto, and N. Ono, J. Appl. Phys., 48, 513 (1977).
40. N. Koizumi, S. Yano, and K. Tsunashima, J. Polym. Sci., B7, 59 (1969).
41. H. Kakutani, J. Polym. Sci., A-2, 8, 1177 (1970).
42. S. Yano, J. Polym. Sci., A-2, 8, 1057 (1970).
43. K. Nakagawa and Y. Ishida, J. Polym. Sci. Phys., Ed., 11, 1503 (1973).

44. A. K. Doolittle, J. Appl. Phys., 22, 1571 (1951); 23, 235 (1977).
45. J. D. Ferry, "Viscoelastic Properties of Polymers", 2nd Ed., John Wiley (1970).
46. K. Matsushige, S. Imada, and T. Takemura, unpublished data.

CRYSTAL NUCLEATION IN FLOWING POLYMER MELTS

Mark S. Pucci and Stephen H. Carr

Department of Materials Science and Engineering
and Materials Research Center
Northwestern University
Evanston, IL   60201

BACKGROUND

The crystallization of polymers either undergoing, or having recently undergone, flow is a means of affecting polymer morphology, and, as such, it is of special interest to the field of polymer processing. This is because virtually all of today's processing methods involve flow of a polymer melt either prior to or during solidification. Since flow is such an integral part of commercial processing, it is important to understand the effect which it can have on crystal texture. With this knowledge, it will not only be possible to control final properties of the material but also it will be possible to avoid inadvertant occurrence of flow-induced crystallization.

The effect of flow on crystallizable polymer systems is not limited to its ability to induce fibrous texture. Flow can also markedly enhance the kinetics of the crystallization process. This is because flow makes the melt-to-crystal transformation more favorable by altering the thermodynamic state of the molten phase. The fact that flow occurs during the processing of polymer melts implies that this enhancement in transformation kinetics becomes the limiting factor in definition of a polymer shaping. Thus, tendency for premature crystallization of the melt rises as processing rates increase. Therefore, in order to attain optimum processing conditions, it is necessary to have as complete a mastery as possible of just how flow is capable of having its effects on crystal nucleation and growth.

Interest in flow-crystallized polymers was first aroused early when it was discovered that a fibrous polyethylene could be precipitated from vigorously stirred solutions [1,2]. This fibrous poly-

ethylene exhibited mechanical properties highly uncommon to quies-
cently crystallized polyethylene [1-3]. Inspection of the fibrous
precipitate under the electron microscope revealed that the polyethylene had crystallized into shish-kebob-like structures [1,2].
This shish-kebob structure consisted of a long central filament of
polymer along which platelets of lamellar crystals were strung.
Electron diffraction studies revealed that the chains in both the
central filament and lamellae were oriented in the direction of the
filament axis [1,4]. Shish-kebobs had also been known to occur under
conditions other than mechanical agitation of a supersaturated
solution [5,6].

     Attention was soon focused on studying the entire morphology of
the shish-kebob more carefully [7-14]. DSC studies performed by
Wunderlich et al. [15], confirmed the fact that molecules of some
degree of extended chain configuration existed in the central filament. This confirmation was based on the fact that the filaments
were thermally stable to temperatures of up to 140°C, i.e. 6 degrees
higher than that for conventional folded chain crystals. Studies
also revealed that the shish-kebobs could be stripped of most of
their lamellar platelet structures when washed in appropriate solvents. This phenomenon of selective dissolution not only attested
to the fact that the central filament was more stable than the
platelet structures, but also indicated that most of the platelet
structures were "overgrowths" of polymer along a more inconspicuous
flow-induced structure [14]. Pennings et al. [10], have suggested
that these overgrowths probably form from polymer which crystallizes
after the flow-induced precipitate is withdrawn from the experiment
vessel.

     Experiments on the crystallization of polymers from their bulk,
either as flowing melts or as strained rubbers were early able to
show parallels with the shish-kebob picture. Specifically, crystallization in strained networks revealed that structures resembling
shish-kebobs had developed when the deformed material crystallized
[16-18]. Similarly, studies seeking to determine whether or not
such structures would form in polymer melts deformed by flow met with
success too. Diffraction patterns from flow-crystallized polyethylene showed that the polymer crystallized in a preferred orientation
[19]. An example of this is seen in Figure 1, which is from the
work of Carr and Fung [20]. Electron micrographs of the oriented
material revealed that the lamellar structures which had formed appeared to be stacked in columns parallel to the flow direction.
Keller and Machin [19] postulated that nuclei similar to the central
filaments of shish-kebob structures were responsible for this colum-
nar morphology. According to Keller and Machin, lamellae would
emanate from these oriented nuclei and grow out in the radial
direction. These lamellae would grow in either a twisted or flat
manner, depending on the magnitude of the flow field. Unfortunately,
electron micrographs of the oriented melt failed to reveal these

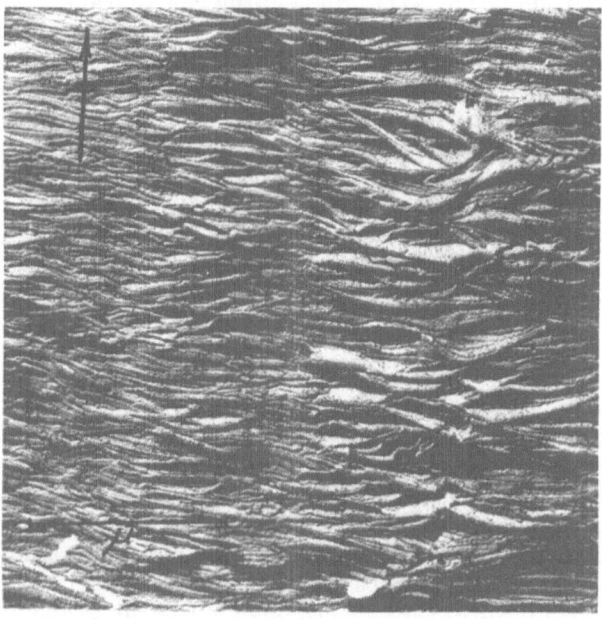

Figure 1:   Transmission electron micrograph of the surface of a
            thin fiber of polyethylene spun from the melt at a high
            elongation rate. Arrow indicates stretching direction.
            The morphology is predominantly comprised of platelet
            lamellae oriented perpendicular to the flow direction
            and seen, therefore, in this micrograph on their edges.
            See reference 20.

nuclei [15,19,21]. This is due to the fact that the vast amount of
lamellar structure appearing in the micrographs makes it difficult
to detect any other structure which may be embedded within it.
However, thermal studies on the oriented solid give good indication
that these filament-like nuclei do exist [22,23].

NUCLEATION KINETICS

     The effect of shearing flow on the induction time in polymer
melts has been studied by many investigators [24-32]. Induction
time is the initial time interval in a phase transformation event
during which no <u>detectable</u> amount of new phase has yet accumulated.
Induction times are inversely proportional to nucleation rate, $\dot{N}$,
provided $\dot{N}$ is not time-dependent. Lagasse and Maxwell [24] have
reported that a critical shear rate must be exceeded before shearing
flows can have an effect on the induction times in polyethylene
melts. Once this critical shear rate was surpassed, the induction

time was found to decrease with increasing shear rate. Lagasse and
Maxwell also found that this critical shear rate was molecular weight
dependent, i.e. higher molecular weight polymers showed smaller
values of critical shear. Haas and Maxwell [25] have also commented
on the existence of a critical shear rate in polybutene-1 systems.
Studies performed by Fritzsche et al. [26,27], indicated that there
also exists a limiting value of shear rate which corresponds to a
saturation of the enhancement effect on nucleation rate. This limiting value of shear rate also appeared to be molecular weight dependent. In this case, polymer of lower molecular weight had higher
values of limiting shear stress.

The effect of molecular weight and temperature on nucleation rate
enhancement at a given shear rate have both been the subject of much
interest [25-27]. Haas and Maxwell [25] report that for low to mod-
erate shear rates, induction times decreased with either an increase
in molecular weight or decrease in temperature. However, for high
shear rates the reverse effect was found to occur, i.e. induction
times increased with either an increase in molecular weight or decrease in temperature. Haas and Maxwell speculated that this anomalous behavior may be the result of one of many factors, including
the possibility that interfacial transport may be disturbed at high
stresses. A very surprising effect which shearing flow has on very
low molecular weight polymer has been observed by Fritzsche et al.
[26,27]. They advanced the idea that when a poly(ethylene oxide)
fraction of molecular weight lower than that needed for entanglement
was deformed at several different shear rates the induction time
increased with increasing shear rate. However, the induction time
under quiescent conditions remained the longest. Fritzsche et al.,
propose that this effect is due to the fact that there is an increased disruption of nuclei with increased shear rate.

So the question now arises: What are the basic laws pertaining
to relationships between the flow imposed on an undercooled polymer
melt and the crystallization that will result? Lagasse and Maxwell
had found that, for moderate-to-high shear rates, the onset of
nucleation occurred at essentially a constant level of strain. It
was also found that the shear rate beyond which the induction strain
became constant was both molecular weight and temperature dependent;
either an increase in molecular weight or decrease in temperature
caused a shift in this limiting shear rate to lower values. The
existence of a constant induction strain had also been proposed by
Krueger and Yeh [28]. However, Tan and Gogos [29] argue that it is
the total energy input per unit volume during the induction period
which is constant, and not the induction strain. (It should be
noted that, for Newtonian fluids, these two parameters are directly
proportional.) Kobayashi and Nagasawa [31] advanced the notion that
it was stretching of chains, caused by the imposed flow, that led
to a reduction in configurational entropy and, in turn, this had the
effect of a free energy elevation. This free energy elevation was

simply additive to the free energy between undercooled melt and
that of the solid which could form from it. Ideas along this line
were discussed earlier by Peterlin [33,34] and were subsequently
expanded upon by McHugh [35]. Demonstration of quantitative agreement between predictions and experimental results was never shown
in those papers.

Meeting the test of good quantitative agreement between observated
kinetics and predictions had been difficult until recent years. One
significant factor is the nature of the nucleus. Most of the papers
cited so far conceive of nuclei that are filamentary in nature. An
original description of this was advanced early by Williamson and
Busse [32]. They proposed that, under the influence of shearing
flow, clusters of molecules in the melt rotate independently of
one another. Molecules, which happen to span between two or more
clusters, experience a stretching deformation due to the rotation of
the clusters. Once sufficient deformation has occurred within the
molecules, nucleation will occur. More recently, Yeh, Hong and
Krueger [36] emphasized that conditions need not lead to filamentary
nuclei; rather, lamellar crystallization could occur simply with an
orientational predisposition of nuclei due to the preferred orientation of chain segments caused by the flow. In all cases, birth of
embryonic nuclei that can subsequently become stable is generally
regarded as occurring by the processes described earlier by Turnbull
[37,38] and later (and in considerable detail) by Hoffman and
Lauritzen [39,40] and by Krigbaum and Roe [41]. It was by building
upon all these ideas that Andersen and Carr [42] were finally able
to show that the desired quantitative agreement between theory and
experiment could be achieved.

The entire time course of transformation from melt to crystalline
solid was studied earlier by Baer and coworkers [43], on quiescent
melts, and their work permits one to recognize how temperature of
crystallization affects the process, either in the presence or the
absence of flow. This is illustrated in Figure 2, where the "quiescent" line tracks the midpoint, $\tau_{1/2}$ (the crystallization "half-time"),
on a transformation curve for different temperatures below some
limiting temperature, the equilibrium melting point (shown by the
vertical dashed lines). Krigbaum and Roe [41], and others, have
emphasized how this limiting temperature will rise with chain orientation developed in the flowing system. Similarly, it is recognized
that increased levels of flow will foreshorten the crystallization
half-times, thus causing the quiescent line to be shifted downwardly
as magnitude of flow rate increases.

MODELLING HOW FLOW ACCELERATES NUCLEATION KINETICS

The special feature of the Anderson and Carr work [42] was the
fraction-wise calculation of free energy elevation due to flow.

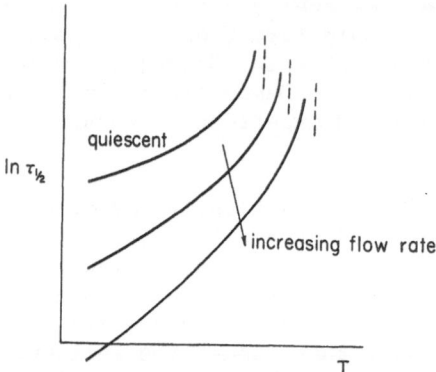

Figure 2:   Schematic representation of how crystallization induction
            time, $\tau_{1/2}$, relates to crystallization temperature, T.
            The effect of flow is seen primarily to depress the
            parabolic curve.

Without that approach, it would have been impossible to produce a
total free energy elevation in the system of the proper magnitude
to agree with experimentally-measured values. One of the interesting conclusions indicated by that work was the fact that only the
chains in the very highest molecular weight fractions are sufficiently stretched by prevailing flow to account for the total free
energy elevation that was observed (see Figure 3). Subsequent work
by Abiodun [44] has extensively investigated the applicability of
this calculation method to many different commercial polyethylene
materials. In all cases, however, the quantity of chains in the
highest molecular weight tail is itself a prediction based on an
accepted method for extrapolating observed molecular weight distributions (from gel permeation chromatograms) to very high values of
molecular weight. One notes that the Adnerson and Carr calculation
method [42] makes a simple assumption that the nuclei born in the
presence of flow have essentially the same character (size, shape
and energetics) as applies to cases concerned with quiescent flow.

   The more recent approach being developed by Pucci [45] seeks to
calculate these free energy elevations with a more realistic set of
models. As with the Anderson work, there is no need for special
assumptions related tothe kind of nuclei which prevail in the onset
of crystallization, but unlike the Anderson work, the prediction of
free energy elevation is done on each of the molecular fractions
present using a force-canonical ensemble (Equation 1) expression
to describe the thermodynamic states (potential energies and entrop-
ies, (see Ref. 46) associated with each of the permissible conformations of the chains in that fraction:

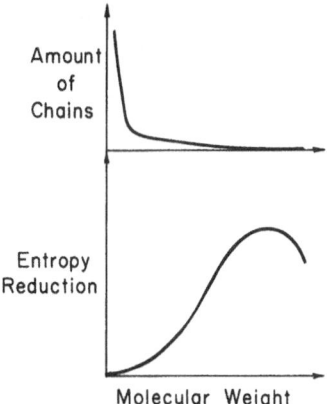

Figure 3:   Schematic representation of the origin of the thermo-
            dynamic driving force for crystallization in a flowing
            melt. The high molecular weight tail of the molecular
            weight distribution is shown in the upper portion of this
            figure, and the amount of entropy reduction contributed
            to the system from each of the values of molecular weight
            is shown in the lower portion. The reason why most of the
            entropy reduction arises from the highest portions of this
            tail, even though the amounts of material involved are
            approaching vanishngly small values, is because of the
            extremely large amount of stretching that these larger
            chains experience in the presence of flow. See reference
            42.

$$\Delta(M,F,T) \;=\; J^M \left(\int_{-a}^{+a} \exp\{F\,x/kT\}dx \right)^M \tag{1}$$

where M is the number of statistical segments in the chain, F is the
force on the chain (simplified as the net result of a force couple
applied only at the ends of that chain), T is temperature, J is the
energy partition function for an individual chain segment, a is the
length of the statistical chain segment, k is Boltzmann's constant,
and x is the distance through which the force, F, elongates the
chain in question. The Gibb's free energy, G, for these chains can
be given as:

$$G = -kT \ln\Delta \tag{2}$$

Combining equations 1 and 2, and differentiating G with respect to
T, yields the following expression for entropy, S:

$$S = Mk \left[1 + T\frac{\delta \ln J}{\delta T} - \theta\coth\theta \right. $$
$$\left. + \ln(2Ja) + \ln(\theta\sinh\theta) \right] \tag{3}$$

where $\theta = \mathrm{Fa}/\mathrm{kT}$. The entropy reduction, ΔS_{flow}, due to the perturb-
ation of a force, F, causes, can be written as:

$$\Delta S_{flow} = -\frac{\xi Ma^2 F^2}{6kT} \qquad (4)$$

where ξ is a monotonically decreasing function of F and is equal to
unity for small values of θ. The challenge now is to employ a meaningful molecular model which will interrelate F and the stretching
it produces. The simplest one chosen is the dumbbell model of
Kramer [47]. If one assigns Gaussian chain properties to the dumb-
bells, equation 4 can be simplified [48]:

$$\Delta S_{flow} = \frac{\mathrm{tr}\underline{\pi}}{2T} \qquad (5)$$

where $\underline{\pi}$ is the extra stress tensor equivalent to the force, F, imposed
on the chain. To take into account the finite extensibility of polymer chains, the Langevin chain is invoked [49]. Once this is done,
the extension of such a dumbbell, x, is obtained:

$$x = \frac{\overline{R_F}}{Ma} = \coth\theta - \frac{1}{\theta} \qquad (6)$$

where $\overline{R_F}$ is the average value of the projection of the end-to-end
vector onto the direction of F.

    Calculation of trace of the stress tensor, $\mathrm{tr}\underline{\pi}$ can be done in a
straightforward manner if the flow field in which the polymer chain
in question exists is that of shear; alternatively, if flow is elongational in character, then a more generalized expression is called
for. For the case of shear, one can use simply rubber elasticity
theroy [50]:

$$\mathrm{tr}\underline{\pi} = P_{11} - P_{22} \qquad (7)$$

where $P_{11} - P_{22}$ is the first normal stress difference in shear. What
one needs to do to obtain some kind of an expression for $\mathrm{tr}\underline{\pi}$ is to
employ a theoretical constitutive equation. The Lodge constitutive
equation [51] has been employed in the Pucci work:

$$\underline{\pi} = \sum_i \frac{E_i}{\lambda_i} \int_{-\infty}^{t} \left(\exp\left\{ -\frac{(t-t')}{\lambda_i} \right\} C^{-1}(t,t') - \delta_{ij} \right) dt' \qquad (8)$$

Here, $\underline{\pi}$ is the summation of terms, each identified by the index, i,
arbitrarily chosen to fit any desired material; their primary contributions to each term are through E_i, an elastic modulus, and λ_i, a
characteristic time constant. Optimal values of the E_i's and

λ_i's are obtained by computer-fitting to dynamic rheological data obtained on the actual material whose crystallization is of interest. The integral weighs each component according to its contribution from the previous stress-strain history that the material had experienced. C^{-1} is the inverse Cauchy strain tensor, describing that history of the deformation imposed on the system. δ_{ij} is the Kronecker delta function. Evaluation of the trace of this stress tensor, which is what is desired when calculating the deformation that models of chain macromolecules experience, results in equation 9:

$$tr\underline{\pi} = \sum_i \frac{2\dot{\varepsilon}}{\nu_i T} \exp\left\{-\frac{t}{\lambda_i}\right\} \int_0^t \exp\left\{\frac{t'}{\lambda_i}\right\} \sigma_i(t')dt' \tag{9}$$

and it can be shown easily from equation 10 that the relative magnitudes of $tr\underline{\pi}$ developed an elongational flow as opposed to values developed from shearing flow are in the ratio of 3 to 1, assuming Trouton's rule applies (a good assumption).

$$\lim_{t\to\infty} tr\underline{\pi} = \sum_i \dot{\varepsilon}\lambda_i\sigma_i \tag{10}$$

As had been indicated above in equation 5, all one needs in order to evaluate the entropic reduction caused by flow is a workable value for $tr\underline{\pi}$, so one now can proceed to predict the effect of this thermodynamic consequence on crystal nucleation kinetics. The approach used here is to take the standard expression for nucleation rate, $\dot{N}$, (40):

$$\dot{N} = N_o \exp\left\{-\frac{\Delta G_t}{kT}\right\} \exp\left\{-\frac{\phi_c}{kT}\right\} \tag{11}$$

where N_o is the pre-exponential term, ΔG_t is the energy barrier for transport of a chain from the melt phase to the crystalline phase, and ϕ_c is $K(\sigma)/\Delta G_x(T)^2$. Here, $K(\sigma)$ is a term dependent on the geometry and surface energetics of a critical-sized nucleus, and $\Delta G_x(T)$ is the Gibbs free energy difference between melt and crystalline phase at a particular (undercooled) temperature, T. The expression for ϕ_c in the presence of flow is, straight-forwardly, expressed as:

$$\phi_{c,flow} = \frac{K(\sigma)}{(\Delta G_x(T) + \Delta G_{flow})^2} \tag{12}$$

ΔG_{flow} is, of course, found by multiplying equation 5 times the absolute temperature. Combining equations 11 and 12 produces the working equation for comparing crystal nucleation kinetics measured in the presence of flow with those measured from the same materials at various temperatures in the absence of flow.

CONFIRMATION OF THEORY

Experimental work has been performed on a variety of hydrogenated
polybutadienes (HPB's) whose unique characteristics are described
elsewhere [52,53]. The first thing that needed to be done was to
obtain values of E_i and λ_i by curve-fitting techniques from dynamic
rheological data; a set of 8 pairs of E_i and λ_i values was arbitrarily chosen (see ref. 45) to fit the data. Once these sets of values
of moduli and time constants had been obtained, it was possible to
prdict free energy, ΔG_{flow}, and these results for a variety of elongation rates can be found in Figure 4. The values of ΔG_{flow} shown
in Figure 4 are plausible values, although their apparent unbounded
rise with time is probably unrealistic. Nevertheless, it can be
seen that at early times (say, 20 seconds), the value of ΔG_{flow} is
already a significant fraction of the total driving force typically
operating during the crystallization of some polymer, like polyethylene.

Direct measurement of crystal nucleation in the presence of
elongational flow has been performed extensively [45] on an instruc-
tive set of hydrogenated polybutadienes, whose identity is seen in
Table I. An elongational viscometer, tailored specifically for use
with the Rheometrics Mechanical Spectrometer and for studies primarily on crystallization effects, has been fabricated [45]. It will
be described in detail elsewhere. An example of the way elongational
viscosity, taken as the instantaneous quotient of tensile stress

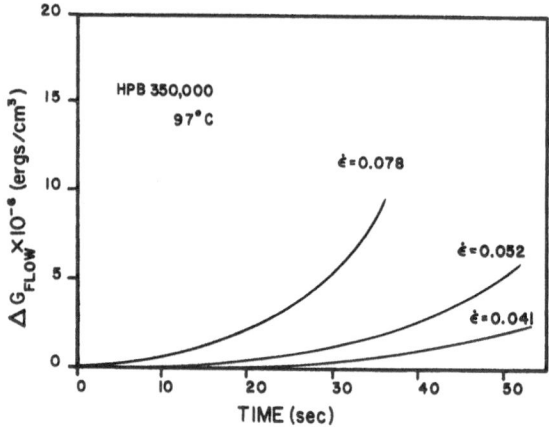

Figure 4:   The buildup with time of free energy due to the amount
            by which the configurational entropy of the chains becomes
            more strongly negative following the imposition of steady-
            state elongation, $\dot{\varepsilon}$. Strain rates are in units of sec^{-1}.
            Values are the result of predictions based on equation 9.
            Results taken from reference 45.

TABLE I

Description of samples used in study

Sample Designation	Description
HPB 350,000L	Linear Polymer $\bar{M}_N = 350,000$
HPB 230,000L	Linear Polymer $\bar{M}_N = 230,000$
HPB 350,000L/95,000L	Blend of Linear Polymers 65% $\bar{M}_N = 350,000$ 35% $\bar{M}_N = 95,000$
HPB 350,000L/130,000S	Blend of Linear and 4-Arm Star Polymer 65% $\bar{M}_N = 350,000$ 35% $\bar{M}_N = 130,000$

divided by (constant) elongation rate, is shown in Figure 5. Here, one sees the effects of two different temperatures on crystallization of a high molecular weight, HPB, being stretched at a relatively low rate. The effect of temperature on the steady-state viscosity (for example, in the interval 20 seconds to 40 seconds) is seen to be very small. Similarly, the first detectable point (identified as the crystallization "induction time") where the rise in viscosity due to the accumulation of newly-born crystals is also hard to find. What is seen as the significantly temperature-dependent aspect is the rate at which viscosity rises after the "induction time" is passed. This apparently is a reflection of a profound temperature-dependence on the crystal growth rate. In Figure 6, one can more easily see differences in the induction time as the elongation rate is changed. These differences between the temperature-dependent effects and the elongation rate-dependent effects have not been systematically reported previously.

When one now takes the induction times obtained experimentally from the flowing melts and calculates the relative increases in nucleation rates using equations 11 and 12 together, one produces the dashed curve seen in Figure 7. One notes that the experimental data seen in Figure 7 is drawn through a set of points obtained from the 350,000 molecular weight HPB, the 230,000 molecular weight HPB, and the blend of linear HPB's. A common, single line was drawn through these data because it represents a constant value of under-cooling (as opposed to the same value of crystallization temperature). The extremely close quantitative agreement between the predicted and

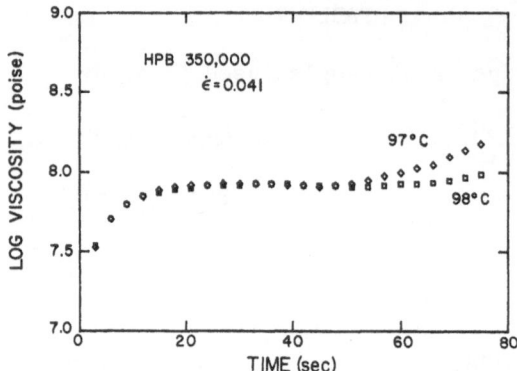

Figure 5:   Tension in the stretching, undercooled strap of HPB mater-
            ial during an elongational crystal nucleation experiment.
            Stress is expressed as viscosity. The effect of larger
            undercooling is seen to be somewhat subtle on what one
            picks for an induction time, but it is more apparent that
            undercooling does affect the growth rate of the crystals
            once they are born. Data from reference 45.

experimental lines is an unprecedented confirmation of the validity
of the model for flow-induced crystallization described in this
paper. It is interesting to note that data for the HPB blend containing four-arm star chains fit the trend rather poorly, with
observed crystal nucleation rates being one-half of what is predicted.
It is speculated that this small, but significant, discrepancy may

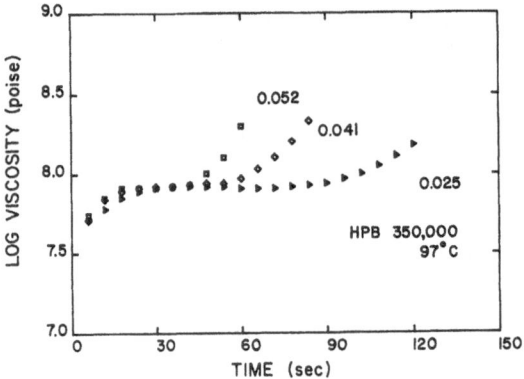

Figure 6:   Tension, expressed as viscosity, in the stretching strap
            of HPB material during a crystal nucleation experiment.
            One can see that elongation rate has a profound effect on
            induction time. Data from reference 45.

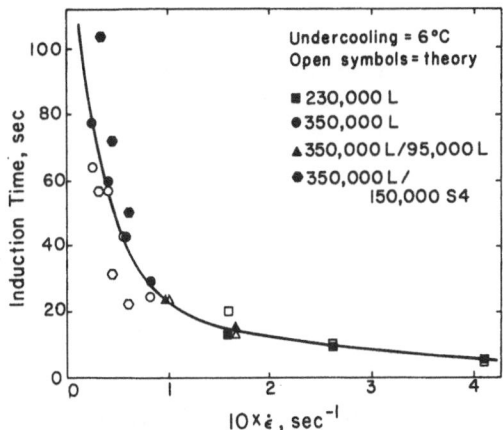

Figure 7:   Master plot for a single value of undercooling (6°C) re-
            lating induction time to elongation rate. The curve is
            a best fit through the data and would intersect the ordi-
            nant at the quiescent crystallization time value (~500
            sec for the linear samples).

reflect the relatively sluggish self-diffusion rate expected for
star-branched chains.

SUMMARY

     Crystal nucleation in undercooled, flowing polymer melts is
accelerated by amounts that can be predicted quantitatively by common expressions for nucleation kinetics. The adaptation of these
relationships culminates in the simple addition of the free energy
difference between melt and solid which could form from it to a free
energy quantity based solely on the reduction in configurational
entropy that flow causes the chains to experience. Precise agreement
between this prediction and experimental results is possible when
one uses a statistical mechanical approach, based on a Lodge model,
to calculate this reduction in configurational entropy.

     Experiments have been performed using a novel elongational vis-
cometer on crystal nucleation kinetics of hydrogenated polybutadienes
(nearly monodisperse model crystallizable polymer). Results obtained
also indicate that acceleration is sensitive to molecular weight to
the >7.5 power and that whichever component of the melt is most
strongly undercooled will cause crystal nucleation kinetics to be
the same as if that component was the only one present.

ACKNOWLEDGMENT

This research was supported by the National Science Foundation through its grant to the Northwestern Materials Research Center (DMR 79-23573).

REFERENCES

1. A. J. Pennings and A. M. Kiel, Kolloid-Z., 205, 160 (1965).
2. A. J. Pennings, "Proc. Inter. Conf. Crystal Growth, Boston, 1966", p. 389, Pergamon Press, Oxford (1966).
3. A.J. Pennings, C. J. H. Schouteten, and A. M. Kiel, J. Poly. Sci., Part C, 38, 267 (1972).
4. T. Kawai, T. Matsumoto, M. Kato and H. Maeda, Kolloid-Z., 222, 1 (1968).
5. D. A. Blackadder and H. M. Schleinitz, Nature, 200, 778 (1963).
6. D. C. Bassett and A. Keller, Phil. Mag., 7, 1553 (1963).
7. A. M. Rijke, J. T. Hunter, and R. D. Flanagan, J. Poly. Sci., Part A-2, 9, 531 (1971).
8. A. Keller and F. M. Willmouth, J. Macromol. Sci., Phys., B6(3), 493 (1972).
9. A. Keller and F. M. Willmouth, J. Macromol. Sci., Phys., B6(3), 539 (1972).
10. A. J. Pennings, A.M.A.A. van der Mark, and A. M. Kiel, Kolloid-Z., 237, 336 (1970).
11. A. G. Wikjord and R. St. John Manley, J. Macromol. Sci., Phy., B2(3), 501 (1968).
12. W. George and P. Tucker, Poly. Eng. Sci., 15(6), 451 (1975).
13. D. Krueger and G.S.-Y. Yeh, J. Macromol. Sci., Phys., B6(3), 431 (1972).
14. F. M. Willmouth, A. Keller, I. M. Ward, and T. Williams, J. Poly. Sci., Part A-2, 6, 1627 (1968).
15. B. Wunderlich, C. M. Cormier, A. Keller, and M. J. Machin, J. Macromol. Sci., Phys., B1(1), 93 (1967).
16. E. H. Andrews, Proc. Roy. Soc. (London), A227, 562 (1964).
17. E. H. Andrews, J. Poly. Sci., A-2 4, 668 (1966).
18. E. H. Andrews and B. Reeve, J. Mater. Sci., 6, 547 (1971).
19. A. Keller and M. J. Machin, J. Macromol. Sci., Phys. B1(1), 41 (1967).
20. P.Y.-F. Fung and S. H. Carr, J. Macromol. Sci., Phys. B6(4), 621 (1972).
21. M. J. Hill and A. Keller, J. Macromol. Sci., Phys. B3(1), 153 (1969).
22. M. J. Hill and A. Keller, J. Macromol. Sci., Phys., B5(3), 591 (1971).
23. T. W. Haas and B. Maxwell, J. Appl. Sci., 14(9), 2407 (1970).
24. R. R. Lagasse and B. Maxwell, Poly. Eng. Sci., 16(3), 189 (1976).
25. T. W. Haas and B. Maxwell, Poly. Eng. Sci., 9(4), 225 (1969).

26. A. K. Fritzsche and F. P. Price, Poly. Eng. Sci., $\underline{14}$(6), 401 (1974).

27. A. K. Fritzsche, F. P. Price, and R. D. Ulrich, Poly. Eng. Sci., $\underline{16}$(3), 182 (1976).

28. D. Krueger and G.S.-Y. Yeh, J. Appl. Phys., $\underline{43}$(11), 4339 (1972).

29. V. Tan and C. Gogos, Poly. Eng. Sci., $\underline{16}$(7), 510 (1976).

30. M. Wolkowicz, J. Poly. Sci., Poly. Symp., $\underline{63}$, 365 (1978)

31. K. Kobayashi and T. Nagasawa, J. Macromol. Sci., Phys., $\underline{B4}$(2), 331 (1970).

32. R. B. Williamson and W.F. Busse, J. Appl. Phys., $\underline{38}$(11), 4187 (1967).

33. A. Peterlin, Makromol. Chem., $\underline{44}$, 338 (1961).

34. A. Peterlin, Pure Appl. Chem., $\underline{12}$, 563 (1966).

35. A. J. McHugh, J. Appl. Poly. Sci., $\underline{19}$, 125 (1975).

36. G.S.-Y. Yeh, K.-Z. Hong, and D. L. Krueger, Polym. Eng. and Sci., $\underline{19}$, 401 (1979).

37. D. Turnbull and J. C. Fisher, J. Chem. Phys., $\underline{17}$, 71 (1949).

38. D. Turnbull, J. Chem. Phys., $\underline{18}$, 198 (1950).

39. J. I. Lauritzen and J. D. Hoffman, J. Res. Nat. Bur. Stand., $\underline{64A}$, 73 (1960).

40. J. D. Hoffman and J. I. Lauritzen, J. Res. Nat. Bur. Stand., $\underline{65A}$, 297 (1961).

41. W. R. Krigbaum and R. J. Roe, J. Poly. Sci., Part A, $\underline{2}$, 4391 (1964).

42. P. G. Andersen and S. H. Carr, Poly. Eng. Sci., $\underline{18}$, 215 (1978).

43. E. Baer, J. R. Collier, and D. R. Carter, SPE Transactions, 5, 22 (1965).

44. O. Abiodun, M. S. Thesis, Dept. of Chemical Engineering, Northwestern Univ., Evanston, IL, Aug. 1981.

45. M. Pucci, Ph.D. Thesis, Dept. of Matls. Sci. & Eng., Northwestern Univ., Evanston, IL, Aug. 1981.

46. T. Hill, An Introduction to Statistical Thermodynamics, Addison-Wesley, Reading, MA, 1960.

47. H. A. Kramers, J. Chem. Phys., $\underline{14}$, 415 (1946).

48. G. Marrucci, Trans. Soc. Rheol., $\underline{16}$, 321 (1972).

49. F. Grun and W. Kuhn, Kolloid Z., $\underline{101}$, 248 (1942).

50. H. Janeschitz-Kriegl, Adv. Polym. Sci., $\underline{6}$, 170 (1969).

51. A. S. Lodge, Elastic Liquids, Academic Press, NY, 1964.

52. W. E. Rochefort, G. G. Smith, H. Rachapudy, V. R. Raju, and W. W. Graessley, J. Polym. Sci., Polym. Phys. Edn., $\underline{17}$, 1197 (1979).

53. H. Rachapudy, G. G. Smith, V. R. Raju, and W. W. Graessley, J. Polym. Sci., Polym. Phys. Edn., $\underline{17}$, 1211 (1979).

FLOW-INDUCED CRYSTALLIZATION AND ORIENTATION FROM THE MELT

J. R. Collier, K. Lakshmanan*, L. Ankrom**,
and S.K. Upadhyayula***

Chemical Engineering Department, Ohio University
Athens, Ohio 45701

ABSTRACT

Highly oriented fibers and ribbons of polyethylene and polypropylene were formed in a process using a single screw plasticating
extruder as the melt source. These extrudates exhibited properties
typical of highly oriented semicrystalline polymers: transparency,
fibrous morphology, elevated melting temperatures, and high initial
and secant moduli and yield strength. This process had demonstrated
the capability of imparting controlled levels of uniaxial and apparently biaxial orientation, that range up to exceptionally high
values. A conditioned polymer melt is fed to specially designed
and operated dies where it experiences elongational flow to impart
orientation in the desired direction(s), and at least the outer
sheath of the extrudate is crystallized in the fixed boundary land
of the die prior to exiting. The flow behavior in one of the dies
used in this process has been modeled illustrating the streamlines
that exist.

INTRODUCTION

This fixed boundary process has demonstrated the capability of
imparting controlled levels of uniaxial and apparently biaxial planar

---

*Current address: Ch.E. Department, Ohio State Univ., Columbus, OH
43210.
**Current address: E.I. DuPont de Nemours & Co., Inc., Parkersburg,
WV 26101
***Current address: National Semiconductor, Santa Clara, CA 95051.

orientation, that range up to exceptionally high values in fiber and
ribbon extrudates (1-6). A single screw plasticating extruder sup-
plies molten polymer to specially designed dies in which the melt is
conditioned, experienced elongation flow to impart orientation in
the desired direction(s), and at least the outer sheath of the extrudate is crystallized in the fixed boundary (i.e., steel walled)
land of the die. The conditioning required apparently involves exposure of the molten polymer to a pressure in the thousands of
pounds per square inch (hundred millions of Pascals) at a temperature
near its atmospheric pressure melting point for at least fifteen
minutes.

Apparently this melt extruder driven process has some similarity to the Solid State Extrusion Process developed by correspondingly
higher moduli values.

As shown in Figure 4, extrudates formed in this process exhibit
a layered or fibrous failure pattern indicative of high chain orientation (2). The scanning electron micrograph reproduced in
Figure 5 further indicates the fibrous fracture nature of the
extrudates, in this case of a polyethylene fiber (1).

EXPERIMENTAL

The die referred to as a uniaxial die had a constant width of
one half inch and a combined depth (i.e., sum of both halves of the
split die) of one sixteenth inch, as shown in Figure 2. The six to
one cross-sectional area reduction was achieved with a straight
walled shaping section as shown. The 12/1 and 2/1 biaxial dies
also had straight walled sides in the shaping section that represented a cross-sectional area reduction of twelve to one, and an
acceleration of flow in the machine and the transverse (i.e., cross-
channel) directions, as shown in Figure 4, in order to also enhance
the transverse direction mechanical properties. Earlier studies
have shown that the exceptional strength enhancement in the machine
direction (extrusion direction) achieved using the uniaxial (constant width ribbon) die is accompanied by easy splitting down the
machine direction. This apparently is a result of the high molecular alignment in the machine direction. The 2/1 biaxial die produced the same dimensioned extrudate as the uniaxial die, whereas
the 12/1 biaxial die produced extrudates that were 1.5 inches wide
and 1/32 inch thick.

A single screw extruder, Brabender model 200, with a barrel
diameter of 0.75 inches and an L/D ratio of 20 was used in this
research. The screw was a nylon type with a compression ratio of
4:1. Between the extruder and the die was a "melt conditioner"
consisting of medium pressure pipe, fittings and a valve. This
unit was rated to withstand 10,000 psi at 200 degrees F, and had

Table 1

POLYMER	DIE TYPE	DIE REDUCTION RATIO	INITIAL MODULUS MACHINE DIRECTION	INITIAL MODULUS TRANSVERSE DIRECTION	TENSILE STRENGTH MACHINE DIRECTION	TENSILE STRENGTH TRANVERSE DIRECTION	ELONGATION TO BREAK	DSC MELTING PEAK ELEVATION	DSC CHANGE IN CRYSTALLINITY UNORIENTED TO ORIENTED
POLYETHYLENE	Uniaxial	6/1	1.6×10^6 psi (11 GPa)	--	--	--	--	10.0C°	--
POLYETHYLENE	Biaxial	12/1	1.9×10^6 psi (13 GPa)	3.6×10^5 psi (2.5 GPa)	4.4×10^4 psi (0.30 GPa)	8.5×10^3 psi (0.059 GPa)	--	10.8C°	61% to 81%
POLYPROPYLENE	Uniaxial	6/1	1.3×10^6 psi (8.8 GPa)	--	4.6×10^4 psi (0.31 GPa)	--	9.5%	8.5C°	--
POLYPROPYLENE	Biaxial	2/1	8.45×10^5 psi (5.8 GPa)	--	--	--	--	8.1C°	24% to 60%

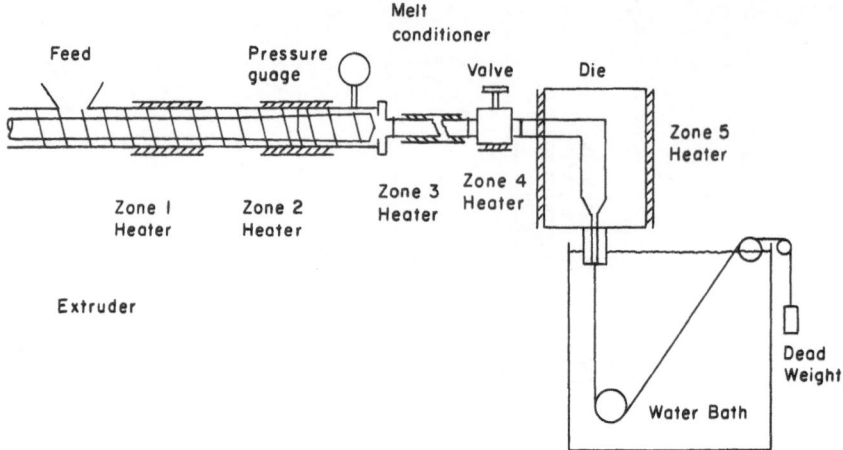

Figure 1.   Schematic Diagram of the Process (without melt
            conditioner).

heaters controlled by Love model 52 on-off controllers. The dimensions of the medium pressure pipes were 1 inch O.D., 0.688 inches
I.D. and a volume of 4.46 cubic inches per foot of length. The die
and melt conditioner were constructed of 316 stainless steel, although the internal surfaces of the split die were coated with a
fluoropolymer dispersion (supplied by DuPont) prior to each run.

     For the samples reported herein, the extrusion was vertically
downward into a water tank and cooled by having the water level 1/4
to 1/2 inch above the tip of the die. A dead weight of 5 pounds
was used as a take up device. Since the ribbons were extruded
vertically downward, they had to transverse a sharp bend around a
guide roller at the bottom of the tank. Other studies in this
laboratory not reported herein were conducted with a cored and in-
ternally cooled die that permitted horizontal extrusion and
eliminated the sharp bends in the ribbon (33).

     Tensile tests were conducted using an Instron Universal Testing
Unit model TT-D with a GR load cell. A Perkin-Elmer differential
scanning calorimeter model DSC 1B was used to measure the melting
characteristics and heats of fusion of the extrudates. For scanning electron microscopy (SEM) samples were attached to the surface

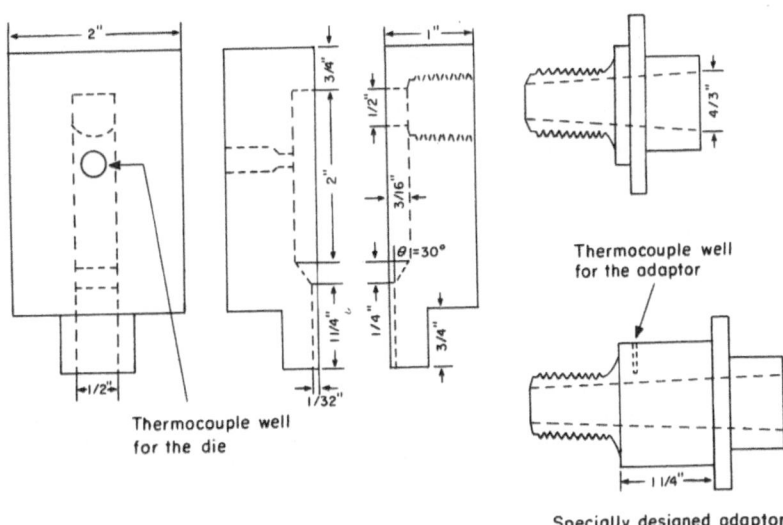

Figure 2.   Schematic Diagram for the Die and the Adaptors (the die
            includes the reservoir and land).

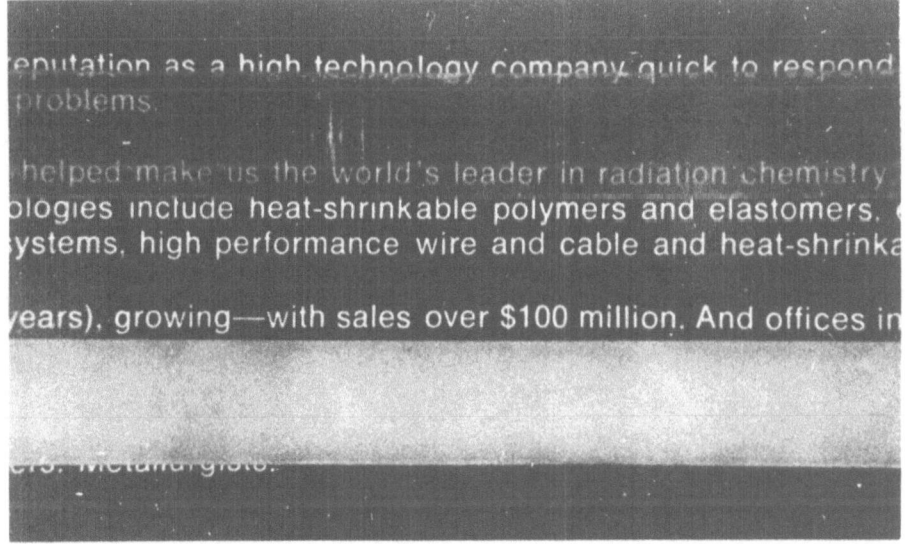

Figure 3.   Comparison of Oriented and Unoriented High Density
            Polyethylene.

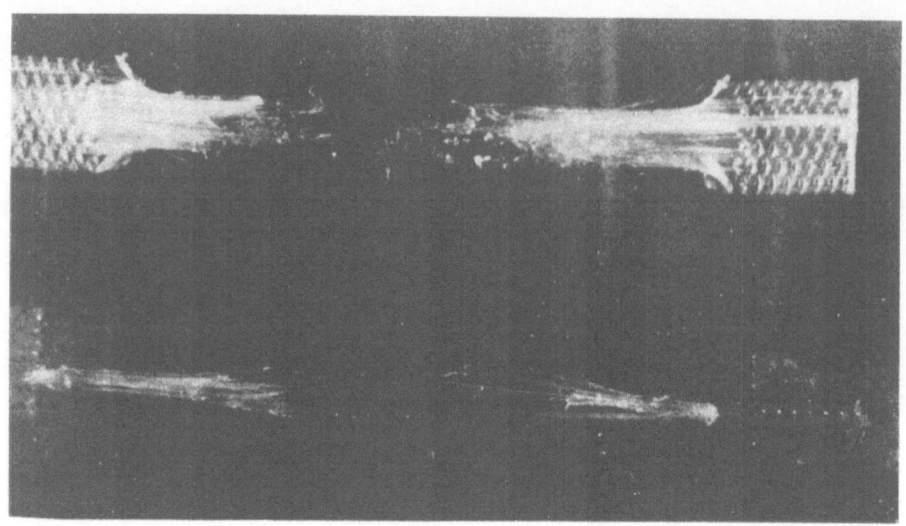

Figure 4.   Tensile Fracture Surface of Polypropylene.

of a one-half inch aluminum pin with silver conducting paint as an
adhesive, then the samples and the holders were covered with gold
in a vacuum evaporator. In this study a Hitachi model HHS-ZR
scanning electron microscope was used.

X-ray characterization using a Warhus camera was graciously
conducted by Dr. Joseph Spruiell of the University of Tennessee.

The polymers used in this study were Alathon* 7030, a high
density polyethylene with a melt index of 2.6, and Marlex** HGZ-
050-02 with a melt index of 5.0 (1,2).

The melt conditioner studies were conducted with two lengths
of pipe, 18 and 28 inches. The lowest extruder discharge pressure
was 1,500 psi at a screw speed of 3 rpm and was varied upward to
6,000 psi at a screw speed of 1.5 rpm. The temperatures in the
five controlled zones; two on the extruder barrel, and one each in
the melt conditioner, valve and die were set for polypropylene at
250, 250, 200, 200, 175 degrees C.

The numerical calculation involved in the modeling portion of
this study was conducted using the IBM 370/158 computer located at
the Ohio University Computer Center.

---

*Registered trademark of E.I. DuPont de Nemours & Co., Inc.
**Registered trademark of Phillips Petroleum Co.

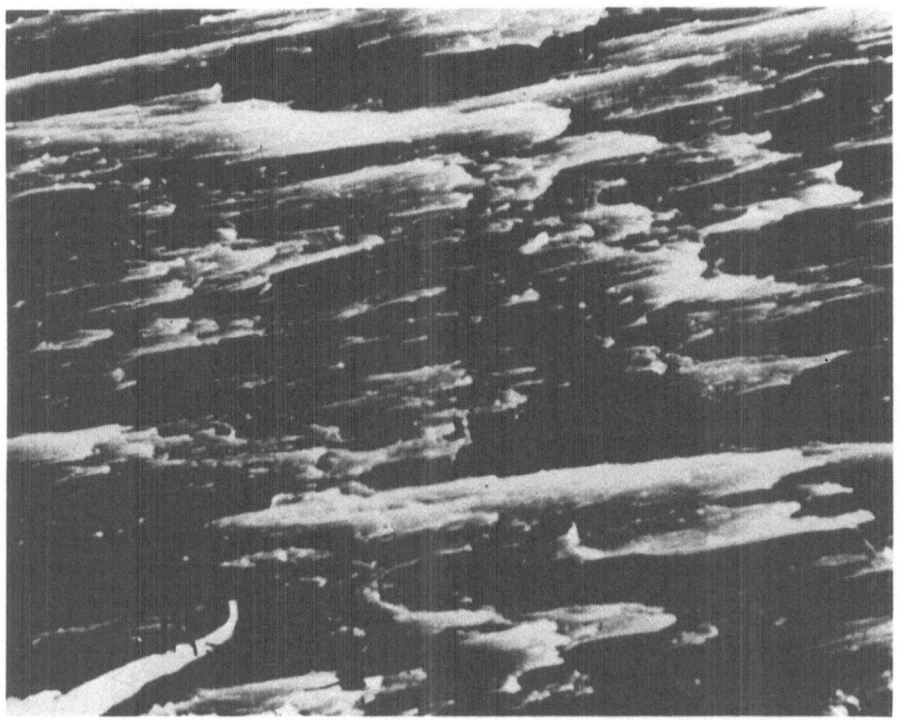

Figure 5. Brittle Fracture of a Polyethylene Sample - 1000X.

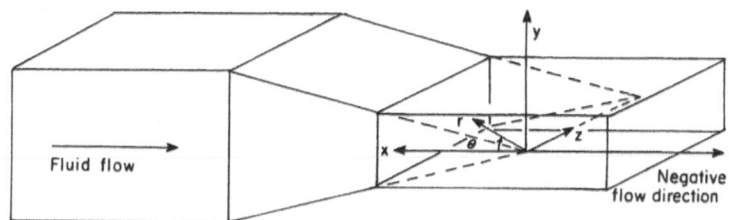

Figure 6. The Coordinate System of the Die.

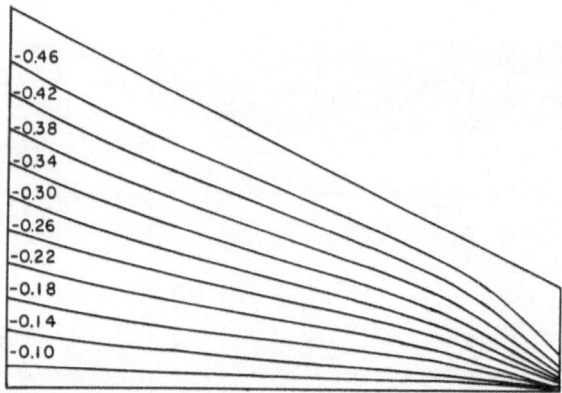

Figure 7.   Stream Function Contour Lines for a Newtonian Fluid in
the Flow Domain.

RESULTS AND DISCUSSION

The coordinate system used for modelling the shaping section
of the uniaxial die is shown schematically in Figure 6. The larger
constant cross-sectional area on the left of this figure is a portion of the reservoir section and the smaller constant cross-
sectional area on the right is the land section. The straight walled
converging section is the shaping section, and this is the only
section modelled. Flow occurs in the negative x direction of the
reservoir and land sections and the negative r direction of the
shaping section. The y coordinate direction in the reservoir and
land section represents the channel height coordinate in those
sections. The z coordinate direction in all three sections of the
die represents the channel width coordinate and is a neutral axis
(i.e., the width is constant and nothing changes with z) in all
three of the die.

The following assumptions concerning the fluid have been made:
1. The fluid is incompressible.
2. The fluid has reached a steady state condition.
3. The fluid is maintained isothermally in the shaping section.

Furthermore, the geometry of the uniaxial die allows further
simplification. The shaping section is symmetrical around the
$\theta = 0$ axis. Hence, only one-half of the shaping section must be
mathematically modeled. The second set of simplifications is due
to the z-dimension remaining constant throughout the shaping section.

Therefore, the velocity in the z-direction is assumed to be equal
to zero while any partial derivate with respect to z also becomes
zero.

Newtonian Fluid

Assuming the fluid is Newtonian causes the continuity equation
and the r and θ components of the momentum balance in dimensionless
variables to become:

$$\frac{\partial v_r'}{\partial r} + \frac{v_r'}{r} + \frac{1}{r}\frac{\partial v'_\theta}{\partial v} = 0 \tag{1}$$

$$v_r'\frac{\partial v_r'}{\partial r'} + \frac{v_\theta'}{r'}\frac{\partial v_r'}{\partial r'} - \frac{(v_R')^2}{r'} = \frac{\partial P'}{\partial r'} + \frac{1}{N_{Re}}$$

$$\frac{v_r'}{r'^2} + \frac{1}{r'}\frac{\partial v_r'}{\partial r'} + \frac{\partial^2 v_r'}{\partial r'^2} + \frac{1}{r'^2}\frac{\partial^2 v_r'}{\partial \theta^2} - \frac{2}{r'^2}\frac{\partial v_\theta'}{\partial \theta} \tag{2}$$

$$v_r'\frac{\partial v_\theta'}{\partial r'} + \frac{v_\theta'}{r'}\frac{\partial v_\theta'}{\partial \theta} + \frac{v_r' v_\theta'}{r'} = -\frac{1}{r'}\frac{\partial P'}{\partial \theta'} + \frac{1}{N_{Re}}$$

$$\frac{\partial^2 v_\theta'}{\partial r'^2} + \frac{1}{r'}\frac{\partial v_\theta}{\partial r'} - \frac{v_\theta}{r'^2} + \frac{1}{r'^2}\frac{\partial^2 v_\theta}{\partial \theta^2} + \frac{2}{r'^2}\frac{\partial v_r'}{\partial \theta} \tag{3}$$

when the dimensionless variables are:

$$4' = r/R \tag{4}$$
$$v_r' = v_r/\bar{v} \tag{5}$$
$$v_\theta' = v_r/\bar{v} \tag{6}$$
$$p' = P/(\rho\bar{v}^2) \tag{7}$$

and $N_{Re} = \frac{\bar{V}R}{\nu}$ is the Reynolds number.

Equation 1 is satisfied by introducing the stream function ψ.

$$v_r' = \frac{1}{r}\frac{\partial \psi}{\partial \theta} \tag{8}$$

$$v_\theta' = -\frac{\partial \psi}{\partial r} \tag{9}$$

Substituting equations 8 and 9 into equations 2 and 3 and eliminat-
ing P in these equations by cross differentiation gives the following equation:

$$\frac{1}{r'}\frac{\partial \psi}{\partial \theta}\frac{\partial \nabla^2\psi}{\partial r'} - \frac{1}{r'}\frac{\partial \psi}{\partial r'}\frac{\partial \nabla^2\psi}{\partial \theta} = \frac{1}{N_{Re}}\nabla^4\psi \tag{10}$$

where

$$A^2 = \frac{\partial^2}{\partial r'^2} + \frac{1}{r'}\frac{\partial}{\partial r'} + \frac{1}{r'^2}\frac{\partial^2}{\partial \theta^2} \tag{11}$$

Equation 10 may be further simplified by using the vorticity relationship, which is the curl of the velocity. The z component of

the vorticity vector (only non-zero component for the uniaxial die) is:

$$\frac{\partial^2 \psi}{\partial r'^2} + \frac{1}{r'^2}\frac{\partial^2 \psi}{\partial \theta^2} + \frac{1}{r'}\frac{\partial \psi}{\partial r'} = -\Omega = \nabla^2 \psi \tag{12}$$

with Ω representing the vorticity vector. Thus equation 10 becomes

$$\frac{1}{N_{Re}}\frac{\partial^2 \Omega}{r'^2} + \frac{1}{N_{Re}r'^2}\frac{\partial^2 \Omega}{\partial \theta^2} + \frac{1}{r'}\frac{\partial \psi}{\partial r'}\frac{\partial \Omega}{\partial \theta} + \frac{1}{N_{Re}} - \frac{\partial \psi}{\partial \theta}\frac{\partial \Omega}{\partial r'} = 0 \tag{13}$$

Equations 12 and 13 are solved numerically to find the two unknowns ψ and Ω.

    In order to develop a numerical scheme for solving equations 12 and 13 it was necessary to specify the ψ value along the boundaries of the flow domain. Several simplifying assumptions were made. First, since the geometry of the shaping section is best described using a cylindrical coordinate system, it is assumed that the flow at the entrance to and exit from the shaping section (two constant value of r) is simple planar Couette flow in the reservoir and land sections. Transforming the reservoir section flow described in cartesian coordinates to cylindrical coordinates and integrating the stream function at the entrance to the flow domain gives:

$$\psi = 2.0\ r'^3 \sin^3\theta - 1.5r'\sin\theta \text{ at } r' = 1.0 \tag{14}$$

The other boundary conditions for the stream function become:

$$\psi = -0.5 \text{ at } \theta = 30^{\circ} \tag{15}$$
$$\psi = 0 \quad \text{ at } \theta = 0^{\circ} \tag{16}$$

The resulting expression for the z component of the vorticity vector is:

$$\Omega = 12.0\ r'\sin\theta \text{ at } r' = 1.0 \tag{17}$$

In addition,

$$\Omega = 0 \text{ at } \theta = 0^{\circ} \tag{18}$$

For this study, the boundary condition for ψ along the wall of the shaping section is assumed to be a constant value as given in equation 15. Since θ has a constant value of 30 degrees along the entire length of the wall, ψ can be a constant value only if ψ is a function of θ not r along the wall. Hence at the wall both the first and second derivates of ψ with respect to r are both equal to zero. Combining this information with equation 12 causes the boundary condition for Ω along the wall to be:

$$\Omega = \frac{-1}{r'^2}\frac{\partial^2 \psi}{\partial \theta^2} \text{ at } \quad \theta = 30^{\circ} \tag{19}$$

    The average velocity, V, and characteristic radius, R, used in this model were (34):

$$\bar{V} = 0.32 \text{ cm/sec}$$
$$R = 0.95 \text{ cm}$$

The value of the kinematic viscosity taken from the literature (35) was

$$\nu = 2.67 \times 10^5 \text{ sec/cm}^2$$

Hence the Reynolds number became 1.14×10^{-7}. The grid sizes were chosen to be:

$$\Delta r' = \frac{0.7696}{60}$$

$$\Delta \theta = \pi/180$$

In addition, the following over-relaxation parameters and smoothing factors were selected. Computational tests determined that these were the optimum values for this model.

$$W\psi = 1.8$$
$$W\Omega = 1.5$$
$$\beta\psi = 0.1$$
$$\beta\Omega = 0.1$$

A contour plot of the stream function is shown in Figure 7. Since the motion is steady the streamlines represent the paths (streaklines) that tracer elements would follow if injected into the flow. The streamlines appear to correspond to lines which radiate from the point of intersection of the nonparallel walls of the shaping section.

Velocity profiles and stress component values were also calculated and plotted but are not included herein, see reference 6 for more details.

Power Law Fluid

The assumptions and simplifications mentioned in the preceding section for Newtonian fluids were also applied to power law fluids. The only basic difference in this section is that the power law relationship replaces the Newtonian constitutive equation. This power law or Ostwalt-de Waele model is:

$$\tau = -M \left| 1/2(\underline{\underline{\Delta}}:\underline{\underline{\Delta}}) \right|^{N-1} \underline{\underline{\Delta}} \tag{20}$$

The parameters M and N vary not only from material to material but also from operating conditions to operating conditions for the same material. The expression in pointed braces in equation 20 is equal to η in equation 21.

$$\underline{\underline{t}} = \eta \underline{\underline{\Delta}} \tag{21}$$

The average velocity, characteristic radius, over-relaxation parameters, smoothing factors, and grid sizes are the same as those given for a Newtonian fluid. A density of 0.86 grams per cubic centimeter was assumed for this non-Newtonian fluid. The power law parameters, M and N, were taken from the literature for low density polyethylene (even though high density polyethylene was used) at 190 degrees C (36) to be the following:

$$N = 1/3$$
$$M = 1.62 \times 10^7 \text{ g}/((\text{cm}) (\text{sec})^{5/3})$$

Figure 8 gives the stream function contour lines for this fluid. Inspection of Figures 7 and 8 reveals that the plots of stream lines are straight for the flow of Newtonian fluids and

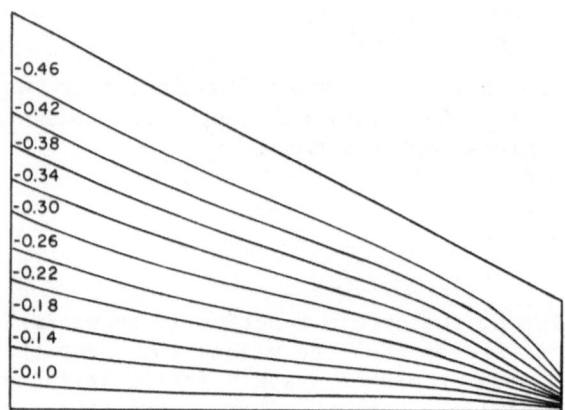

Figure 8.   Stream Function Contour Lines for a Non-Newtonian Fluid
in the Flow Domain.

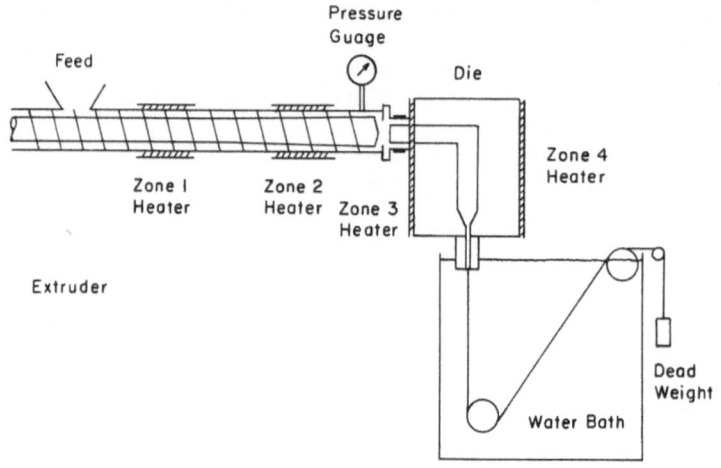

Figure 9.   Schematic Diagram of the Process (with melt conditioner).

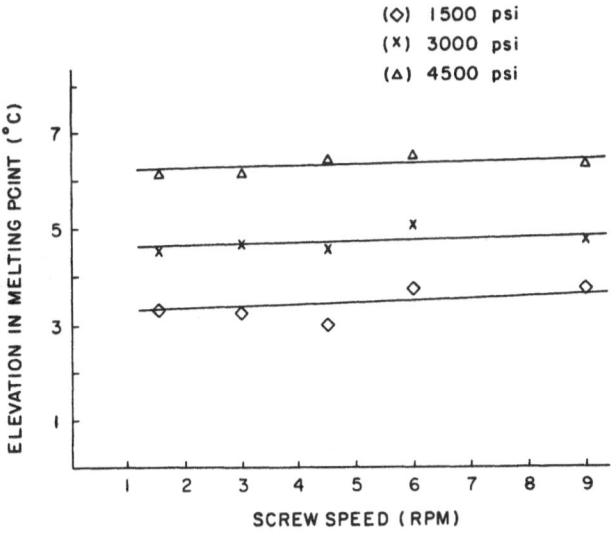

Figure 10. Screw Speed versus Melting Point Elevation.

slightly curved for power law fluids in this domain. This probably
results since the stream function, ψ, is only affected by changes
in the "viscosity" term through the vorticity term. Hence, if the
vorticity term is not the dominant term, the stream function will
be relatively unaffected by changes in viscosity dependence. This
means that the stream function equations and vorticity equations for
Newtonian and power law fluids were only weakly coupled in this
study. Both Newtonian and power law models represent only viscous
fluids. It is quite possible the elastic effects in viscoelastic
fluid models, that more accurately reflect the behavior of the
polymer in this process, could cause stronger coupling between the
equations. If stronger coupling did result then the stream lines
should be different for different fluid models.

Melt Conditioning
       As discussed in the experimental section of this paper, a "Melt
Conditioner" consisting of medium pressure pipe and fittings was
inserted between the extruder discharge and the die, as shown in
Figure 9. Previous studies in this laboratory had indicated that
the polymer melt had to be "conditioned" in the die reservoir prior
to being subjected to an elongational flow in the shaping section.
This was necessary in order to achieve the transparent, high
strength exudates. The conditioning consisted of at least fifteen
minutes residence at a pressure of 1,500 to 6,500 psi (the only
range investigated) and near the atmospheric pressure melting point.
Longer times apparently had no effect. Once the polymer conditioned

in the die reservoir was purged by fresh polymer melt from the
extruder, the pressure decreased and the extrudate was converted to
a normal translucent, lower strength ribbon. The melt conditioner
should provide suffcent additional residence time in the melt at
the proper temperature and pressure conditions to provide continuous
conditioning.

The elevation of the DSC melting peak compared to a normal
extrusion or a recrystallized samples has been shown in previous
studies (1,2) to be a good indication of successful operation of the
process. As shown in Figure 10, the elevation in DSC melting peak
temperatures apparently indicates that the conditioner is providing
continuous conditioning. The DSC melting peak temperature elevation
did not change over a period of four hours of operation. The data
in Figure 10 also suggest that the process is dependent upon the
extrusion discharge pressure but is relatively insensitive to the
extruder screw speed.

Planar Orientation

Ribbons extruded using the uniaxial die, shown in Figure 2,
that exhibit high strength and transparency also characteristically
exhibit weak transverse strength, probably due to the high chain
alignment with the machine (extrusion) direction. In an attempt
to improve the transverse strength of the extruded ribbons, dies
were designed that should provide elongation flow in both the
machine and transverse directions. One of these "Biaxial" dies is
shown in Figure 11. Detailed study of this die will indicate that
basic design criterion for the die was to provide an equal acceleration in the machine and transverse directions. This was accomplished by proper decrease in the cross-sectional area of the flow
channel while simultaneously increasing the width of that channel.
The high degree of orientation reported earlier (1,2) for uniaxially
oriented ribbons and fibers was maintained in extrudates formed in
the biaxial dies. Furthermore, the low transverse strength exhibited by uniaxially oriented ribbons was not only eliminated, the
transverse strength was approximately double that exhibited by
unoriented ribbons.

Polyethylene ribbons were extruded using the die, illustrated
in Figure 11, that had a cross-sectional area reduction of 12/1;
and polypropylene ribbons were extruded through a similar die with
a reduction ratio of 2/1. The polyethylene ribbons exhibited a
moduli of elasticity in the machine and transverse directions of
13 and 2.5 GPa (1.9×10^6 psi and 3.63×10^5 psi) respectively;
whereas the polypropylene ribbons had a machine direction modulus
of 5.8 GPa (8.42×10^5 psi). Compared to unoriented extrudates
the increase in moduli for polyethylene was 1,000% in the machine
and 100% in the transverse direction; whereas for polypropylene
the corresponding increase in the machine direction was 100%.
Polypropylene had a lower increase due to a much lower cross-

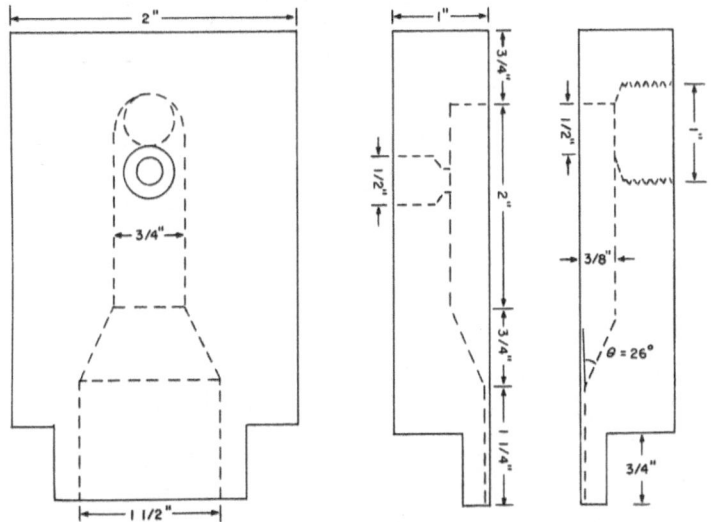

Figure 11.   Schematic Diagram of the 12/1 Biaxial Die.

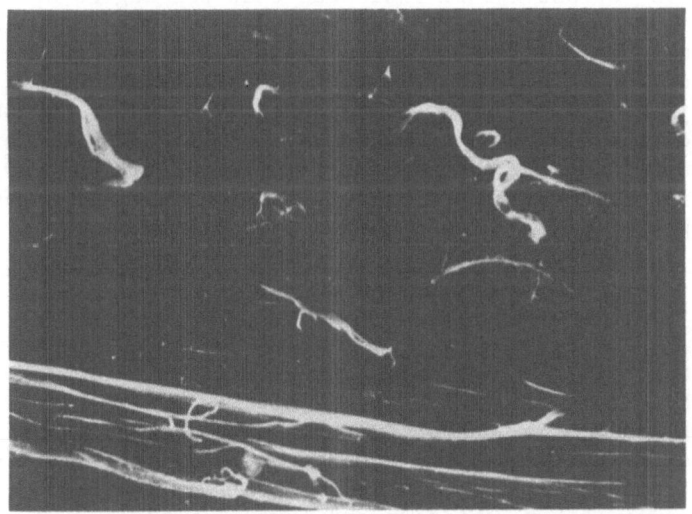

Figure 12.   Scanning Electron Micrograph of Polypropylene.   Extruded
Using the Die Shown in Fig. 11 - 1000X.

sectional area reduction ratio in the shaping section. Transverse
directional strength was not measured on polypropylene samples due
to their narrow width. The failure mechanism in both the machine
and transverse directions had a fibrous layered appearance similar
to the machine direction failure pattern indicated in Figures 4 and
5 for uniaxially oriented fibers and films.

Figure 12 is the reproduction of a 1,000X scanning electron
micrograph of the interior of a polypropylene ribbon. This micrograph suggests that major fibrils lie in the machine direction and
minor fibrils in the transverse direction. Differential scanning
calorimetry revealed single elevated melting peaks of the biaxial
extrudates of polyethylene and polypropylene to be 8.1 and 10.8
degrees C, respectively, higher than unoriented samples.

The results of x-ray analysis performed by Dr. Spruiell (37)
and partially shown in Figure 13 did not indicate biaxial orientation of the polymer chain axis. Apparently the 200 crystal direction tends to align with the normal direction of the ribbon (perpendicular to the ribbon surface), and the 001 (or chain axis direction) tends to align with the machine direction. The patterns also
suggest that there is a slightly greater tilt of the c-axis toward
the transverse direction (ribbon width direction) than toward the
normal direction.

These x-ray results are consistent with current modeling
studies. This current effort indicates that the degree of orientation imposed by dies such as the biaxial dies will be dependent
upon both the acceleration in each direction and upon the velocity
in each direction prior to entering the shaping section. The acceleration in both directions was designed to be equal but the
reservoir section velocity in the transverse direction is zero and
is significant in the machine direction. Therefore, the current
modeling effort is giving results as least qualitatively consistent
with the experimental x-ray and morphological results.

CONCLUSIONS

In this paper, the flow induced crystallization and orientation process that produces high strength transparent extrudates and
maintains die land shape was discussed. Mathematical modeling of
this process indicates that the non-Newtonian characteristics of the
polymer melt studied do not significantly affect the flow streamlines. Streamlines for Newtonian and power law fluid were nearly
identical. Elastic effects of viscoelastic fluids that were not
considered herein could be significant.

The melt extruder driven process discussed in this paper can
be run in a continuous fashion if sufficient residence time occurs

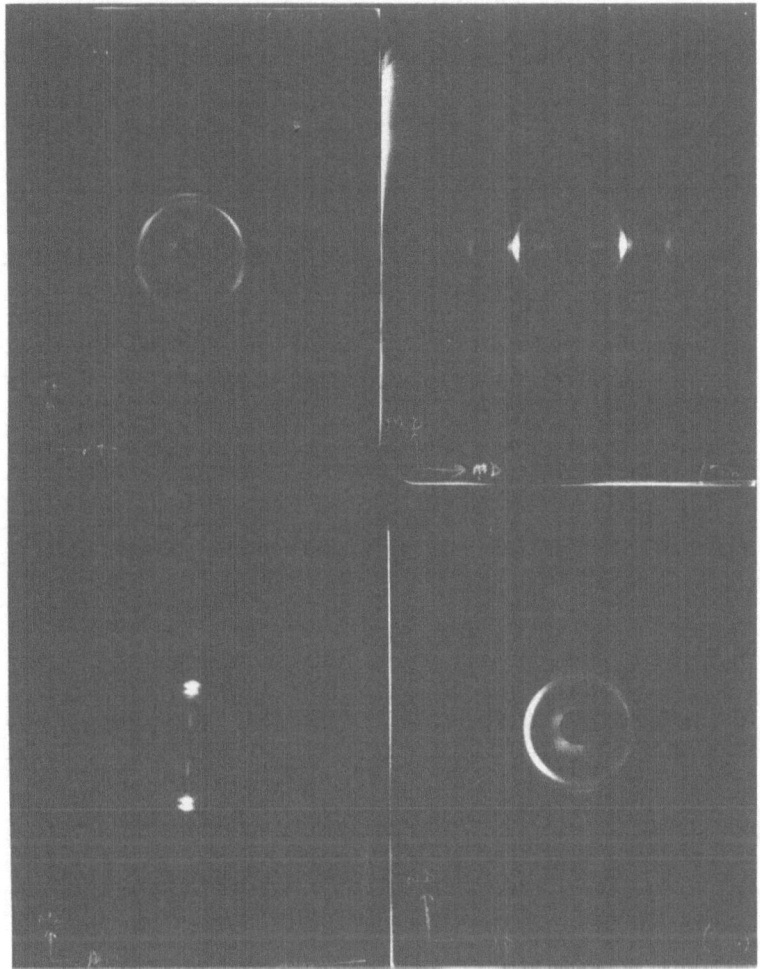

Figure 13. X-Ray Patterns for Polypropylene. Extruded Using the
Die Shown in Fig. 11.

at pressures of 1,500 to 4,500 psi or higher, and temperatures
near the atmospheric pressure melting point. Apparently at least
fifteen minutes is necessary and longer time is not required.

Planar orientation in the machine and transverse directions can
be imparted by expanding the width of the flow channel in the
shaping section while the cross-sectional area is being reduced.
Whether or not true biaxial orientation is imparted has not yet

been determined, however, the transverse direction strength has been
markedly improved while maintaining the exceptional machine direction
strength, transparency and elevated melting points of the extrudates
compared to the previously reported uniaxially oriented ribbons and
films.

ACKNOWLEDGMENTS

    Financial support from the National Science Foundation through
grants ENG76-17592 and CPE79-12376, and from the Monsanto Textile
Fibers Co. enabled this work to be completed. Samples were supplied by the E.I. DuPont de Nemours & Co., Inc. and by the Phillips
Chemical Co. X-ray analysis was conducted by Dr. Joseph Spruiell
of the University of Tennessee.

REFERENCES

1. J.R. Collier, T.Y.T. Tam, J. Newcome, and N. Dinos, Polym. Eng.
    Sci., 16, 204 (1976).
2. J.R. Collier, S.L. Chang, S.K. Upadhyayula, Midland Macro.
    Mono., No. 6, edited by R.L. Miller (1978).
3. J.R. Collier and S.K. Upadyayula, A.I.Ch.E. National Annual
    Convention, Miami Beach, Nov. 15, 1978.
4. S.K. Upadhyayula, Thesis, Chem. Eng., Ohio University (1978).
5. K. Lakshmanan, Thesis, Chem. Eng., Ohio University (1981).
6. L. Ankrom, Thesis, Chem. Eng., Ohio University (1981).
7. A. Keller and M.J. Machin, J. Macromol. Sci.-Phys., B1, 41
    (1969).
8. M.J. Hill and A. Keller, J. Macromol. Sci.-Phys., B3, 153
    (1969).
9. E.H. Andrews, Proc. Roy. Soc. (London), A277, 562 (1964).
10. T.W. Haas and B. Maxwell, Polym. Eng. Sci., 9, 255 (1969).
11. A.J. Pennings and A.M. Kiel, Kolloid-Z.Z. Polymere, 205, 160
    (1965).
12. K. Kobayashi and T. Nagasawa, J. Macromol. Sci.-Phys., B4,
    331 (1970).
13. D. Krueger and G.S.Y. Yeh, J. Appl. Phys., 43, 4339 (1972).
14. T. Kawai, R. Kamoto, K. Ehara, T. Matsumoto, and H. Maeda,
    Sen-i Gakkaishi, 26, 80 (1970).
15. A.K. Fritzsche and F.P. Price, Polym. Eng. Sci., 14, 401 (1974).
16. B. Wunderlich, J. Polym. Sci., A2, 3697 (1964).
17. B. Wunderlich and T. Davidson, J. Polym. Sci., Pt. A-2, 7,
    2043 (1969).
18. T. Davidson and B. Wunderlich, J. Polym. Sci., Pt. A-2, 7,
    2051 (1969).
19. C.L. Gruner, B. Wunderlich and R.C. Bopp, J. Polym. Sci., Pt.
    A-2, 7, 2061 (1969).

20. R.B. Prime and B. Wunderlich, J. Polym. Sci., Pt. A-2, 7, 2061 (1969).
21. E.S. Clark and L.S. Scott, Polym. Eng. Sci., 14, 682 (1974).
22. J.R. Collier, T.Y.T. Tam, J. Newcome, and N. Dinos, Polym. Eng. Sci., 16, 204 (1976).
23. R.S. Porter and A.E. Zachariades, International Symposium on Mechanical Properties of Crystalline Polymers, U. of Mass., Amherst, MA, Oct. 1980.
24. J.R. Kastelic, ACS National Meeting, Atlanta, GA, April, 1981.
25. S. Goldstein (ed.), "Modern Developments in Fluid Dynamics", Vol. I, p. 105, Oxford University (1938).
26. P.N. Kaloni, J. of Phys. Soc. of Japan, 20, 132 (1965).
27. N.C.P. Ramacharyula, Zeitschrift fuer Angewandte Mathematik und Mechnik, 47, 9 (1967).
28. C.D. Han and L.H. Drexler, J. Appl. Polym. Sci., 17, 2369 (1973).
29. W.J.R. Chen, Ph.D. Dissertation, Chem. Eng., Syracuse Univ. (1971).
30. D. Greenspan, J. Fluid Mech., 57, 167 (1973).
31. L. Lapidus, "Digital Computation for Chemical Engineers", McGraw-Hill Book Co., Inc., NY (1962).
32. B. Appelt, L. Wang and R.S. Porter, Society of Rheology 52nd Annual Meeting, Williamsburg, VA, Feb. 1981.
33. C.B. Rao, Thesis, Chem. Eng., Ohio University (1980).
34. S.L. Chang, Thesis, Chem. Eng., Ohio University (1977).
35. J. Brandrup and E.H. Immergut (Editors), "Polymer Handbook", Second Edition, John Wiley and Sons, NY, pp. v-21 (1975).
36. S. Middleman, "Fundamentals of Polymer Processing", McGraw-Hill Book Co., NY (1977).
37. Personal Communication, Dr. Joseph Spruiell, Univ. of Tennessee, Dec. 1978.

HYDROSTATIC EXTRUSION OF GLASS REINFORCED

AND UNREINFORCED CELCON® POM

J. Kastelic, A. Buckley, P. Hope,† and I. Ward†

Celanese Research Company
Summit, New Jersey 07901
†The University of Leeds
Leeds, LS2 9JT, United Kingdom

ABSTRACT

    Several of the solid state shaping methods developed for metals
have been found applicable to polymers. When applied to plastics such
methods as rolling, drawing and hydrostatic extrusion can impart substantial molecular orientation, consequently enhancing physical
properties. Of these techniques, hydrostatic extrusion may have advantages in that glass filled resins may be employed and more complex
profiles may be produced, including hollow sections. Here, we investigate the processing speeds and physical properties attainable by
hydrostatic extrusion of glass reinforced and unreinforced Celcon®
polyoxymethylene. Large size scale and modest 2-10 area reductions
are explored so that commercially feasible production rates can be
approached. Both resins can be successfully processed by hydrostatic
extrusion. Furthermore, at these modest reductions, significant gains
in physical properties are achieved. The strongest beneficial effects
are found in tensile and impact strengths. Surprisingly, the
elongation of the glass filled resin is also improved.

INTRODUCTION

    Many solid state shaping techniques have been developed for use
with metals and several of these have been found applicable to
polymers. Rolling, drawing and hydrostatic extrusion are such
examples. These shaping methods may offer certain advantages, listed
in Table I, over conventional melt processing.

175

Table I. Possible Advantages of Solid State Processing

Ordinary and intractable polymers
Longer fibers and higher loadings
No size or thickness restrictions
Speed advantage
Lower processing temperature
Potential for physical property enhancement

Of great interest is the possibility for property enhancement. These techniques, when applied to polymers, are known to impart substantial chain orientation in the machine direction and consequently lead to greatly improved physical properties [1-12]. Values for the tensile modulus of polyethylene in the range of 10^6 to 10^7 psi have been frequently achieved. Furthermore, other physical characteristics of the material can be enhanced as well. Reported improvements are listed in Table II.

Rolling and drawing are now used to a limited extent commercially primarily in the production of specialized products such as strapping tape and high modulus fiber. To our knowledge, hydrostatic extrusion (H.E.) is not in commercial use with polymers although it is a routine profiling technique with ductile metals such as aluminum, tin and lead.

As a shaping technique to extend the property range of ordinary and, possibly, intractable polymers, H.E. offers several advantages over rolling and drawing. First, there are few constraints on

Table II. Diverse Properties Favorably Effected by
Hydrostatic Extrusion

Significantly increased transparency with crystalline
    polymers
Somewhat elevated melting point
Lower permeability and diffusion rate
Enhanced chemical resistance
Thermal conductivity significantly increased along the
    extrusion direction
Increased resistance to indentation and wear

---

profiles. Accurate rod, sheet and structural shapes such as I-beam
and channel and hollow shapes such as tube and pipe have all been
demonstrated [13,14]. Secondly, these shapes are all feasible on a
larger size scale than possible with rolling and drawing [7]. The
latter do not scale up readily. Finally, it is known that H.E. is
applicable to several fiber reinforced resin systems. Thus, there is
the possible added benefit of reinforcement fiber alignment. Height-
ened ductility and fracture energy have been found to be imparted in
this instance [1,12].

Here we investigate H.E. as a processing method for reinforced
and unreinforced polyoxymethylene. The technical and practical limits
of the process are explored at commercially feasible size scale and
speed ranges. Our approach is to produce oriented materials from
resins of interest by H.E. and evaluate properties as a function of
processing conditions. Resins to be reported on here include glass
reinforced and unfilled Celcon® grades. Parameters to be varied include processing temperature, extrusion pressure and speed, and the
reduction ratio. Attention is on identifying commercially feasible
processing ranges as defined by property enhancement and production
rate criteria.

EXPERIMENTAL

The types of equipment which can be used for the H.E. of polymeric materials and the basic operating procedures have been previously described in the literature [7]. The work reported here has
been done with two of the basic machine types. In-house work at small
size scale was done on a piston-ram device while all large scale extrusions were performed with the pressurized vessel system at the
University of Leeds, U.K.

The Leeds equipment can handle billets up to 3.25 inches in
diameter and about 12 inches long. Extrusion is in a horizontal
direction and a winch system is available which can exert a considerable pull on the emerging extrudate and assist the hydrostatic forces.
For the work reported here, only a small load was applied for the sole
purpose of keeping the extrudate straight.

Large sale extrusions (University of Leeds work) were performed
at 150°C with a castor oil pressure fluid. The dies used were
straight wall conical entry with a half angle of about 30°C. The
transition from the cone to the land region was generously rounded
since this has been found to reduce extrusion instability problems.
Most extrusions were produced using a one-inch die. The billets were
machined to the appropriate diameters to give the desired reduction
ratios which ranged in this work from 2 up to 10. Extrusion speeds
were determined as a function of applied pressue at each ratio chosen.

Physical properties were measured as a function of reduction ratio and this evaluation is ongoing.

Small scale in-house work was carried out on a converted Instron Capillary Rheometer. This instrument is normally used for melt viscosity determinations but does have the necessary configuration and pressure capability for small scale hydrostatic extrusion. When used for H.E., the conventional dies are replaced with trumpet entrance cone dies of much larger final diameter and shorter land length than that employed for viscosity measurements. One-eighth inch diameter circular profile rod dies were used for all the in-house work reported here.

A wide range of processing conditions were evaluated in-house. Temperatures from room temperature to 150°C and production speeds from 1.2 to 120cm/minute were explored. Most work was done at a single reduction ratio, 6, which is in the middle of the range explored at Leeds. The influence of processing temperature and speed on physical properties were determined from in-house specimens while that of reduction ratio was investigated with Leeds' specimens. A silicon oil pressure medium was utilized for in-house extrusion.

Due to the round cross section of all extrudates, physical testing proved to be cumbersome. The one-inch rod extrusions from Leeds were sliced and machined into flat bars 1/8 inch thick by 1/2 inch wide and about 4 inches long. These were used directly for flex tests and trimmed into dog bones for tensile tests. In-house specimens were tested without machining. However, only tensile and flexural moduli and flexural strength were evaluated.

Work reported here was performed on Celanese Celcon® M25-01 unreinforced and GC-25A glass reinforced resins. The latter contains 25% chopped glass. These materials were obtained as melt processed rod. In-house work required billet stock which would fit the Instron barrel. The standard .375-inch diameter rods received were centerless ground about 10 mil undersize to avoid seizure in the bore while at operating temperature. Billets for the University of Leeds were machined from 3-inch or 1-1/2 inch outside diameter melt extruded rod from the same supplier. Some problems with visible central defects, including bubbles, cracks, and porosity, were experienced with the larger rod but enough stock of adequate quality was obtained so that all extrusions could be accomplished.

RESULTS AND DISCUSSION

Production Speeds and Pressures

Literature suggests that production speeds are extremely low for hydrostatic extrusion, e.g. a few millimeters per minute. The natural

implication is that processing costs for H.E. on a per pound basis are
high. However, it must be realized that the goal of much of the published work is the achievement of noteworthy physical properties. Our
own performance aims are intentionally set somewhat lower so that the
processing speed situation may improve. The upper limits to extrusion
speed have not been previously investigated, particularly at low and
intermediate reduction ratios.

Speed-pressure relationships are mapped out for all the reduction
ratios and resins studied in Figure 1. This is for large scale
extrusion. As expected, lower reduction ratios require lower driving
pressure. Also, there is only a slight slope to most of the curves
meaning that not too much of a premium is paid in pressure in order to
speed up the process. This is particularly true for the lower
reduction ratios and the unfilled resins.

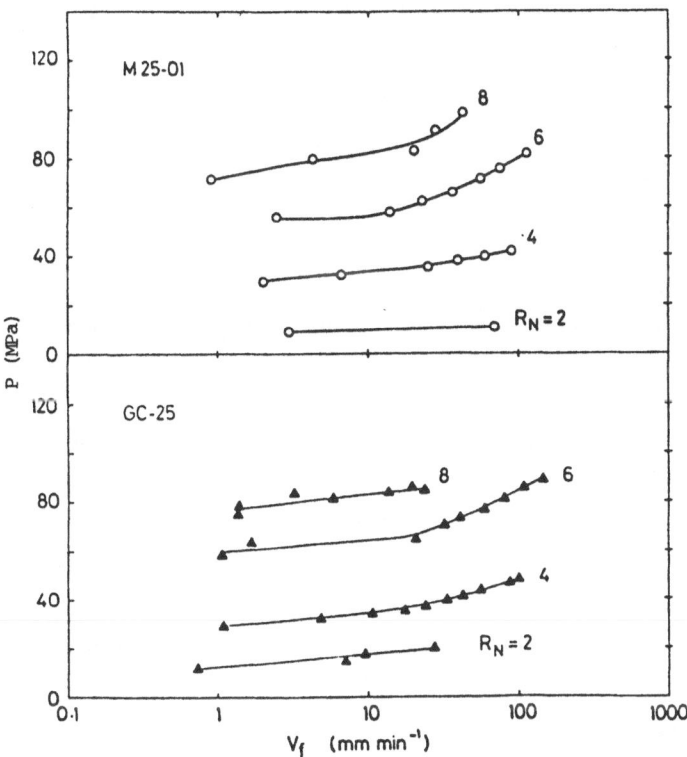

Fig. 1. The relationship of extrusion velocity to applied hydro-
         static pressure in the production of large 1" diameter solid
         Celcon® rods (hydrostatic extrusion performed at 150°C).

It appears that, at 6X and lower, production rates in excess of 100mm/minute are feasible for all these resins. At 8X, there is evidence of a sharp upturn in the required pressure as speeds approach 100mm/minute. Experience has shown that this upturn signals the approach of an adiabatic heating region in which partial intermittent melting of the extrudate occurs. The mechanical work dissipated at the die in plastic deformation and friction is the origin of the heating. If the speed is pushed much beyond this point, the process becomes unstable and violently erratic. The resulting extrudate becomes very irregular in cross section and, of course, loses physical properties.

In preliminary work at an extrusion ratio of 10X, the region of instability was found to begin at very low processing speed - below 10mm/minute. This is commercially very unattractive so further experiments at 10X were not performed. At reduction ratios below 6X, there is no evidence for any speed limitation; thus, it is very likely that production speed at this scale could be pushed well beyond 100mm/minute.

The relation between reduction ratio and the required driving pressure for M25 and GC25 resins is plotted in Figure 2. This is for slow speed (10mm/min), large scale extrusion. The relationship is surprisingly linear for both resins and although the fiber reinforced resin requires slightly higher pressures, the pressure difference between them is not great. Their molecular weights are different and this fact may compensate for the filler. Straight line curve fitting can readily be accomplished with the data and equations for both resins are given in the figure.

The processing temperature has a very strong influence on the necessary extrusion pressure. Data for small scale extrusion at a reduction of 6 and 20mm/minute processing speed are plotted in Figure 3. As the processing temperature is decreased, the pressure must be increased. This holds true for both the filled and unfilled Celcon® and is a highly linear relationship. The curves may continue to lower temperatures and higher pressures but this exceeds our present experimental capabilities.

The effect of size scale on extrusion pressure can be noted by comparing the overlapping data of Figures 2 and 3. For a reduction ratio of six to one at an extrusion temperature of 150°C it can be seen that extrusion at large size scale requires measurably less pressure than at small scale. The argument has been previously made that friction effects become significant in small scale extrusion [15] and, thus, raise the required operating pressures. This seems to be borne out here. It is interesting to note that the pressure increase needed for glass-filled resin at small scale is greater than that required for the unfilled. This is consistent with the higher coef-

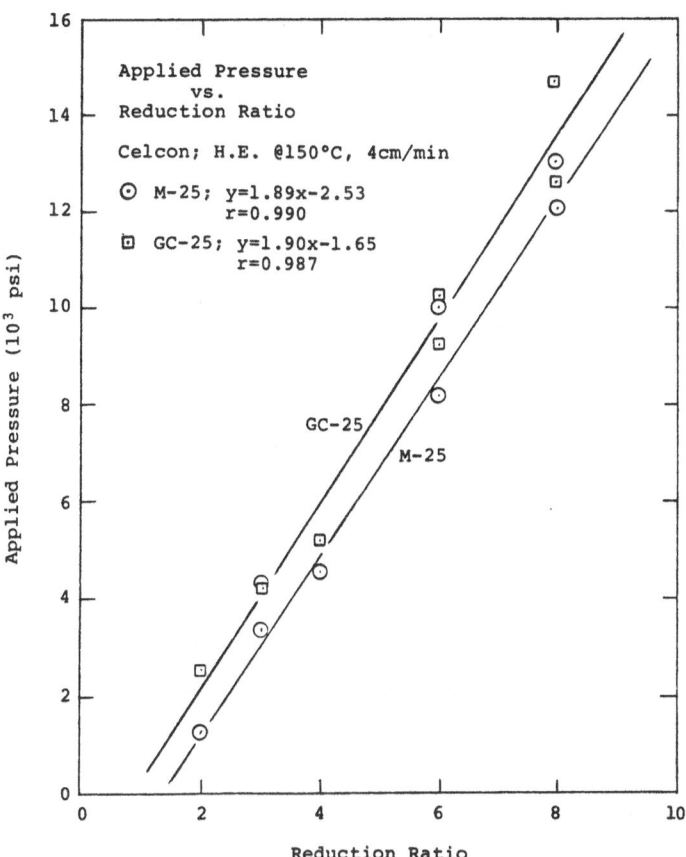

Fig. 2. The applied pressure required to extrude Celcon® resins at
         150°C and 4cm/min extrusion speed plotted as a function of
         area reduction ratio. Linear curve fit equations are given
         in legend.

ficient of friction of the filled resin. It is the chosen reduction
ratio and extrusion temperature which primarily dictate the required
extrusion pressure. Resin molecular weight, filler content, size
scale and process speed have a much lesser effect on pressure. It
must be pointed out that the pressures which must be employed for the
H.E. of polymers are well within the pressure capabilities of the
available production machinery now used for metals.

Physical Properties of Celcon® Extrudates

     The processing temperature chosen for H.E. can have some bearing
on the physical properties achieved. The tensile moduli measured on
small scale extrusions of draw ratio 6 are plotted in Figure 4 as a

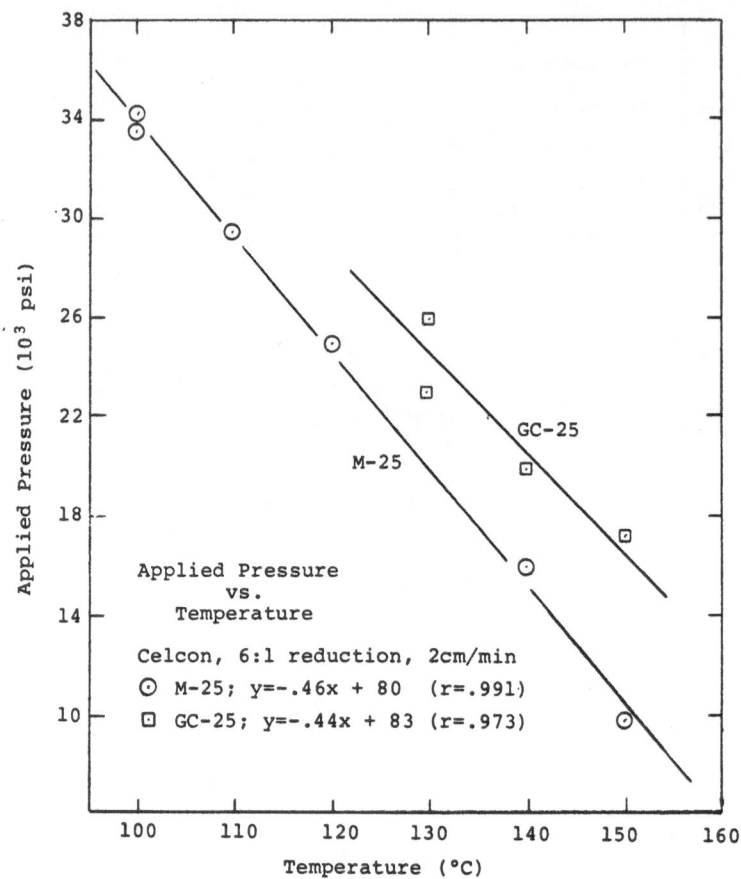

Fig. 3. The pressure required as a function of temperature to sustain
         extrusion at 2 cm/min, through a trumpet taper, 6:1, round
         die. The data is for filled and unfilled Celcon® hydrostat-
         ically extruded on lab scale equipment. Linear curve fit
         equations are given for each resin in the figure.

function of this temperature. For the M-25 resin, there is only a
slight effect. The tensile modulus increases slightly as the temperature is reduced. This effect is easily accounted for. Die swell is
known to be reduced at lower processing temperature and the corresponding real reduction and orientation are thus improved. This
results in slightly increased properties but it does come at a price.
Recall that the extrusion pressure is significantly increased by lowering the temperature. Barring pressure limitations, the M-25 can be
processed from 150°C down to at least 100°C and possibly lower. There
is a slight property advantage by processing at lower temperatures.
This does not appear to be true for the GC-25 resin. There seems to

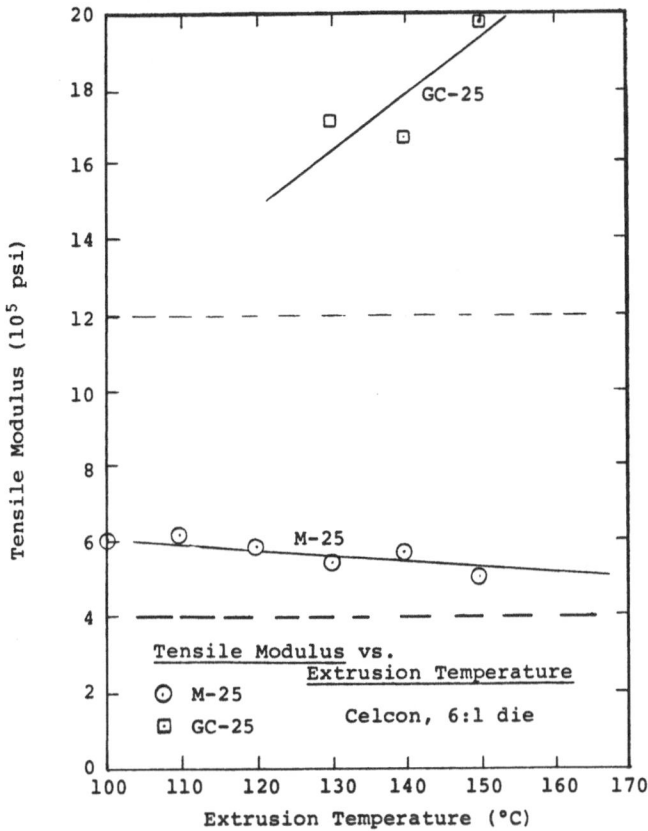

Fig. 4. The effect of extrusion temperature on the resulting tensile
         modulus. Data is for filled and unfilled Celcon® resins
         hydrostatically extruded on a lab size scale at an area
         reduction ratio of 6. Baseline properties of melt processed
         materials are indicated by dashed lines.

be a deleterious effect in lowering the processing temperature from
150°C. The tensile modulus falls though more data is needed. This
could be due to glass fiber breakage under possible increased mechanical forces encountered during H.E. at the lower temperatures.
Another possibility is fiber-matrix bond damage. At the higher processing temperatures the bond could be tougher, subjected to lower
mechanical stresses or even reformable if of simple physical nature.
As a consequence, the recommended H.E. processing temperature for both
optimum properties and reduced extrusion pressure, in the case of the
glass-filled resin, is in the vicinity of 150°C. This is the
temperature at which all large scale work was conducted.

The actual physical properties observed are a function of the amount of reduction. The tensile and flexural moduli achieved by large scale H.E. at different reduction ratios are presented in Figures 5 and 6, respectively. The melt process values are indicated on both figures for comparison.

For the M-25 resin there are percentage gains in these properties ranging as high as 80%. The relationship with reduction is roughly linear for both moduli and the curves extrapolate back to approximately the melt process value at zero reduction ratio. Deviation from the extrapolated curve is seen at very low reduction, particularly for the flexural modulus of the GC-25 resin. Here, the data suggest a

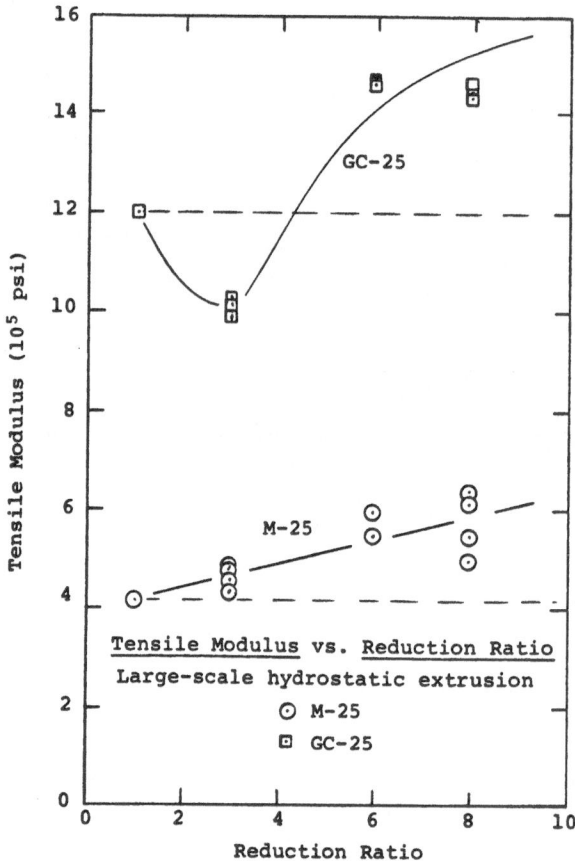

Fig. 5. Tensile modulus of hydrostatically extruded Celcon® as a function of reduction ratio. Large scale extrusion at 150°C. Melt process values indicated by dashed lines.

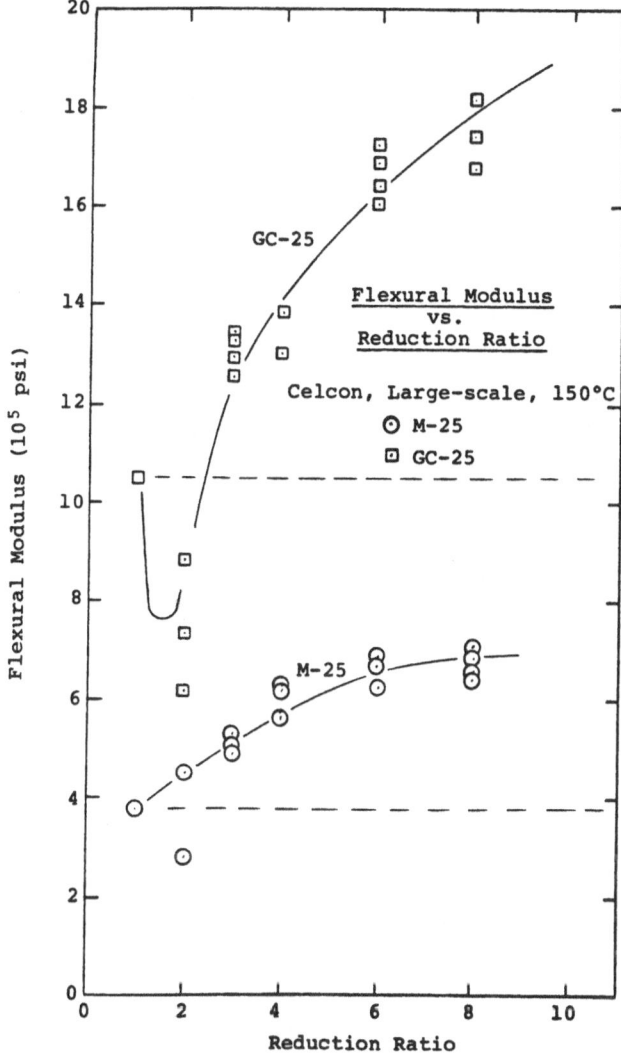

Fig. 6. Flexural modulus plotted against reduction ratio. Large
         scale hydrostatic extrusion of filled and unfilled Celcon®
         resins.

marked reduction in these properties to below the melt process values.
Causes for such a decrease are not clear. This could be evidence for
a more reversible draw process taking place at low reduction which
could lower the observed moduli of the extrudates. The die swell does
surprisingly increase below a draw ratio of three for both of the
resins.

Breaking Strength

The breaking strengths of both the reinforced and the unreinforced resins are greatly improved by H.E. The effect has been attributed to orientation and defect removal. As can be seen in the
data, Figure 7, the relationship is linear with reduction ratio.
Surprisingly, the unreinforced resin achieves higher strengths than
the reinforced. Improvements of up to 400% are possible with the

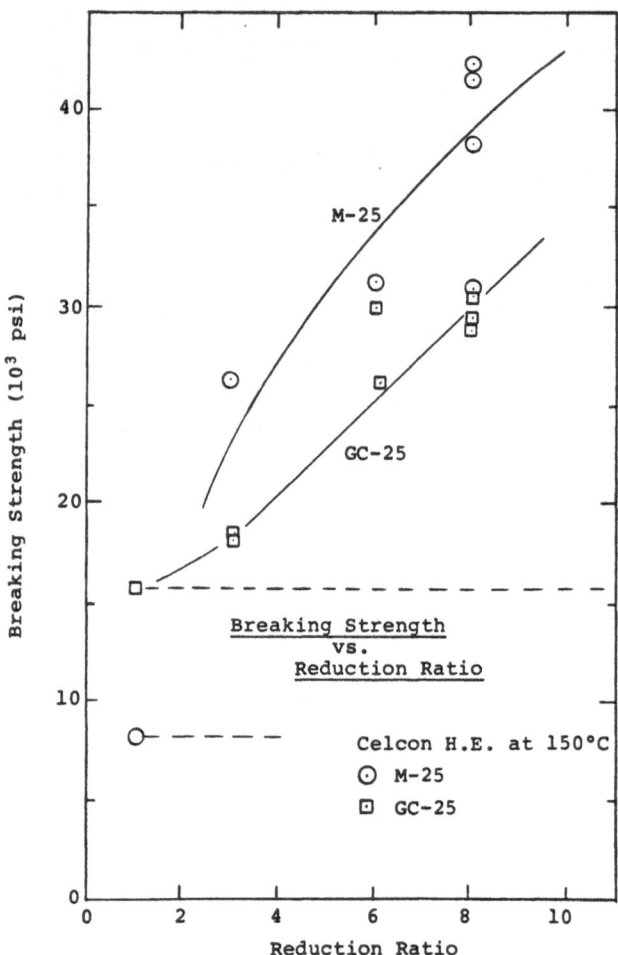

Fig. 7.   Breaking strength as a function of reduction ratio for large
          scale hydrostatic extrusion of Celcon® resins. Extrusion
          performed at 150°C. Breaking strengths of melt processed
          resins indicated by dashed lines.

former. This is the most significant property effect observed at low reduction with H.E. and it could serve as an important factor for uses of this processing method.

Elongation and Impact Properties

There are appreciable effects of orientation by H.E. upon the elongational and impact properties of both filled and unfilled Celcon® resins. For the unreinforced resin, the elongation to failure, Figure 8, is not greatly effected by H.E. to low reduction but there is a strong fall-off at higher reduction. This loss is not as great as might be expected. M-25 resin reduced by a ratio of eight to one still retains 15-20% elongation.

The situation is remarkably different with the glass reinforced resin. The elongation is actually increased by H.E. processing as observed by Buckley and Cassin several years ago [1-12]. Elongations as high as 12% can be achieved. The effect is most pronounced at low reduction and falls somewhat at higher reductions. Still, at a reduction of eight to one, about 6% elongation is retained. These are respectively high values for a glass-filled resin.

This is a surprising phenomenon since most uniaxially oriented composites fail at the fiber elongation which, for normal reinforcing fibers, is only a few percent or less. The enhanced ductility of the GC-25 resin here suggests that some composite mechanism for elongation has been made possible after H.E. This could involve fiber slip. An alternate explanation is that fiber lengths have been greatly reduced so that the resin is behaving as a particulate filled system. Only weak evidence for the latter has been found in our studies of fiber length distribution.

These effects on elongation are reflected in notched Izod impact performance. Unfortunately, tests of these materials are usually invalidated by incomplete failure. The samples develop multiple splits in the extrusion direction and then are able to flex away from the impact hammer. A value of 8.3 ft-lbs. per inch of notch was successfully measured for the GS-25 resin of reduction 3. The other samples in this extrusion series are believed to have even higher values. Clearly, enhanced ductility and impact strength are property effects that could make H.E. a valuable processing technique.

Potential Commercial Processing Range

Work to date suggests that there are ranges in processing conditions within which the H.E. of Celcon® resins can be practiced without difficulty. The extrusion temperature is one parameter with limits. For the M-25 resin, temperatures from about 150°C down to 100°C and possibly lower are all acceptable. However, for the glass-filled

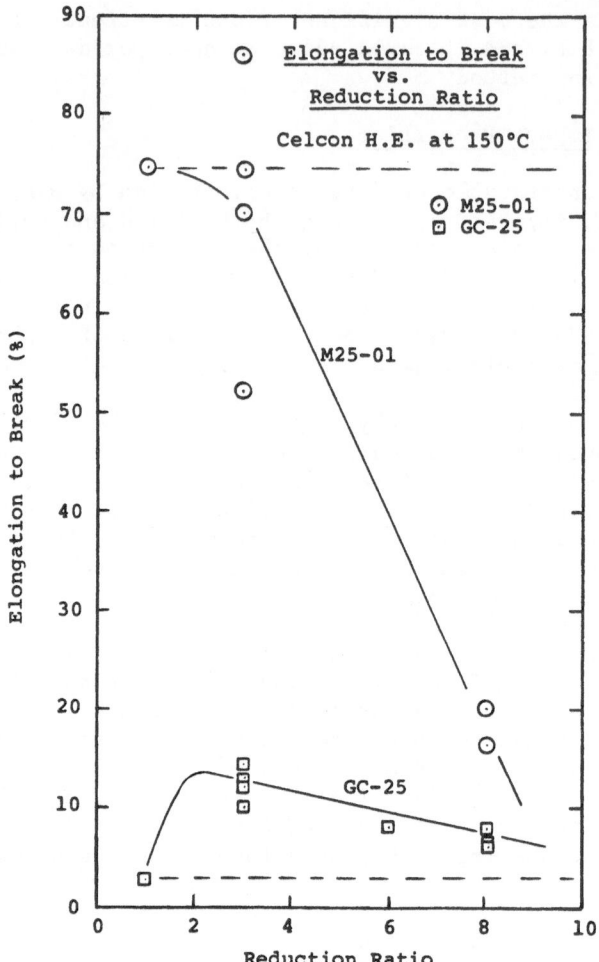

Fig. 8. Elongation to break vs. area reduction ratio for Celcon®
resins hydrostatically extruded at 150°C. Large scale
extrusions performed at Leeds. Melt processed values for
both resins are indicated by dashed lines.

GS-25 resin the range is more constricted. Processing via H.E. should
take place at about 150°C. Any higher and adiabatic heating can too
easily put the resin above its melting point. Any lower and the
mechanical properties appear to fall.

Reduction ratios are also restricted to within practical limits.
Above about eight, the process is stable only for small scale
extrusion. Even at small scale the speed must be somewhat reduced,

consequently throughput drops. At a reduction of eight, large scale
extrusion is stable but several types of defects can arise; with M-25
surface cracks routinely appear (Figure 9). This problem might be
solved by specially designed dies but no work of this type has yet
been attempted. The cracks do not interfere with the present goal of
physical property evaluation. In addition, voids and pores in the
starting stock seem to originate annular blemishes in the products.
Better quality starting material made by careful melt processing could
eliminate this.

Mechanical property improvement is the main feature which recom-
mends H.E. Table III summarizes properties which are readily achievable with Celcon® resins. The reduction ratio here is six.
Generally, all the machine direction properties are improved. One
exception is the elongation of the unreinforced resin which does drop
somewhat. The most significant increases occur in the tensile
strengths. The improvement can range as high as several hundred
percent. Izod impact and, for glassreinforced resin, the elongation
are also markedly improved.

The trade-off, of course, is in the transverse properties. These
have not been thoroughly investigated yet. A sheet die has been made
at Leeds and materials produced with it should make possible accurate

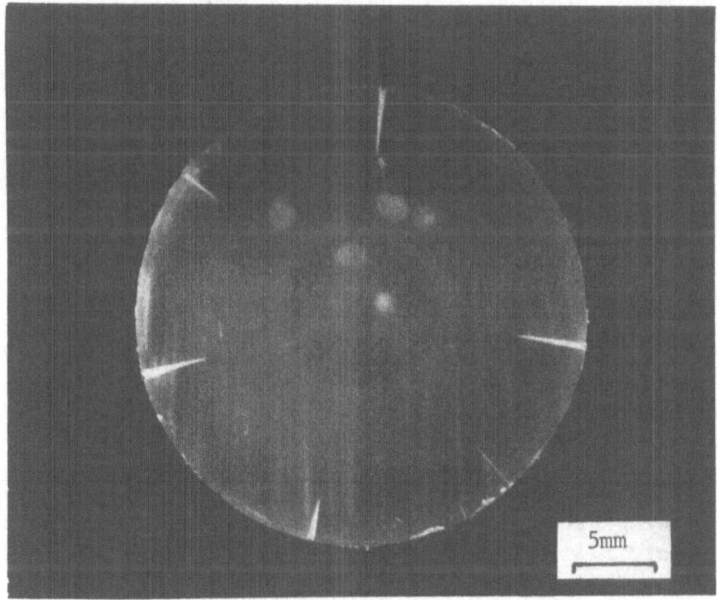

Fig. 9.   Cross-sections of unfilled Celcon® (M-25) hydrostatically
          extruded to a reduction ratio of 8. Cracks and circular
          blemishes are present. 2.5X magnification, back lighting.

Table III. Physical Properties of Hydrostatically Extruded Celcon®
vs. Melt Processed

		M-25		GC-25	
		Melt Processed	6X	Melt Processed	6X
Tensile Strength	10^3 psi	8.8	31	18	26
Tensile Modulus	10^5 psi	4.1	5.8	12.0	14.5
Elongation	%	60	40	2-3	9
Flexural Modulus	10^5 psi	3.75	6.5	10.5	16.5
Izod	Ft-lbs/in	1.3	>8	1.1	>8

transverse property determinations. Preliminary tests on tubing suggest that transverse properties are seriously affected. For tubing of
reduction six, a transverse modulus of 2.2×10^5 psi, a tensile
strength of about 5000 psi and an elongation of only 2.2% were
measured. Clearly, the mechanical anisotropy characteristic of
materials processed by H.E. must be considered in selecting end-uses.
Future work with biaxial dies, crosslinkable resins and radiation
crosslinking might prove to eliminate this reservation.

It is clear that melt processing for ordinary requirements and
composite methods for high strength application will always be strong
competition for any of the solid state shaping methods. However, as
higher and higher performance is demanded from processes and products,
hydrostatic extrusion may become attractive for many specialized
shaping applications.

REFERENCES

1. A. Buckley and H.A. Long, Polym. Eng. & Sci., 9:115, 1969.
2. J. M. Alexander and P. J. H. Wormell, Annals of the C.I.R.P.,
    19:21, 1971.
3. K. Imada, T. Yamamoto, K. Shigematsu and M. Takayanagi, J. Mat.
    Sci., 6:537, 1971.
4. T. Williams, J. Mat. Sci., 8:59, 1973.
5. H. W. Starkweather Jr., T. F. Jordan and G. B. Dunnington, Polym.
    Eng. & Sci., 14:678, 1974.
6. E.S. Clark and L. S. Scott, Polym. Eng. & Sci., 14:682, 1974.
7. "Ultra-high Modulus Polymers", ed. by A. Ciferri and I. M. Ward.
    Applied Science Pub., London, 1977, Chapts. 1 & 2.
8. K. Nakayama and H. Kanetsuma, J. Appl. Polym. Sci., 23:2543,
    1979.

9. H. N. Yoon, K. D. Pae and J. A. Sauer, Polym. Eng. & Sci., 16:567, 1976.

10. S. M. Aharoni and J. P. Sibilia, J. Appl. Polym. Sci., 23:133, 1979.

11. W. T. Mead, C. R. Desper and R. S. Porter, J. Polym. Sci. Phys., 17:859, 1979.

12. A. Buckley and C. Cassin, U.S. Patent 3,642,976, 1972.

13. D. M. Bigg, E. G. Smith, M. M. Epstein and R. J. Fiorentino, Polym. Eng. & Sci., 18:908, 1978.

14. P. S. Hope, A. G. Gibson, B. Parsons and I. M. Ward, Polym. Eng. & Sci., 20:540, 1980.

15. A. G. Gibson and I. M. Ward, J. Polym. Sci., Phys., 16:2015, 1978.

THE INTERACTION OF LIQUID ENVIRONMENTS

WITH HARD ELASTIC HIGH IMPACT POLYSTYRENE

Kim Walton, Abdelsamie Moet, and Eric Baer

Department of Macromolecular Science
Case Institute of Technology
Case Western Reserve University
Cleveland, Ohio 44106

INTRODUCTION

The hard-elastic behavior exhibited by specially prepared polymers has been under intensive investigation since its patenting in the mid-1960's [1]. Recently, the phenomenology of hard elasticity in crystalline polymers has been thoroughly reviewed by Sprague [2] and more recently, Cannon, McKenna and Statton [3]. Common mechanical and physical properties of these unique materials include (a) an initial Hookean elasticity, (b) high recoverability from large strains, (c) energetic retractive forces, (d) rehealing after work softening, and (e) constant cross-sectional area during deformation.

These properties were once thought to be unique to crystalline polymers which were melt spun and annealed under uniaxial tension, producing a morphology of rows of lamellar islands stacked parallel to one another, separated by interconnecting fibrils. Several models have been proposed to explain the mechanical and physical behavior, most of which was based on one mechanism or another within the lamellae.

We have recently reported, however, that by inducing profuse crazes in high-impact polystyrene (HIPS) via uniaxial tension [4,5], or in polycarbonate via sinusoidal tensile loading [6], materials were produced which exhibited all properties associated with hard elastic crystalline polymers. We attribute the mechanical behavior to deformation-related changes in surface energy via a subfibrillation process of the craze fibrils [5]. The breakdown of major craze fibrils into minor fibrils has been shown in thin films of atactic polystyrene by Beahan et al. [7]. Their illustration of the process is depicted in Fig. 1.

193

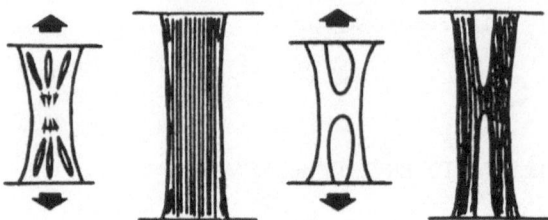

Figure 1: The breakdown of major craze fibrils into subfibrils as
depicted by P. Beahan et al.

The cyclic stress-strain behavior of HIPS, precrazed at 40% elongation (Fig. 2), typifies the mechanical behavior of hard elastic
materials. The initial elastic modulus reflects the composite nature
of the system which is ideally composed of thin layers of solid material bridged by extended fibers. We propose that at a critical stress,
"yield" occurs signalling a complex process of sequential subfibrillation and orientation. The stress plateau is believed to indicate

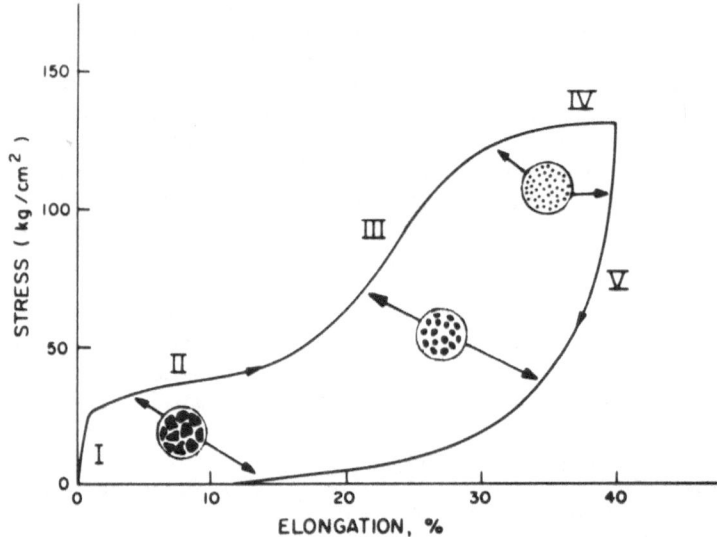

Figure 2: Typical hard elastic behavior of precrazed HIPS in association with the sequential subfibrillation process.

the incorporation of new fibrillar material into the craze from the
solid boundary. An important feature of this mechanism is the revers-
ibility of the sequential fibrillation process upon load reversal.
This can be understood in light of the fact that attraction due to
secondary forces (e.g. van der Waals) is sufficient to produce adhesive joints, between polymers, of strength equal to that of the polymers themselves. Freshly formed microfibrils remain within very
close proximity and reformation of fibrils is perhaps feasible.

It should be emphasized here that similar mechanical behavior is
exhibited by other materials with different superstructures and, in
most cases, are due to entirely different mechanisms. Examples include wool and keratin [8] and synthetic block copolymers with domains
of hard and soft segments such as styrene-isoprene [7] and styrene-
butadiene [10]. The novelty, therefore, does not lie in the discovery
of a new mechanical behavior but rather, in the establishment of a
specific structure-property relationship. Our consensus is that polymeric materials, irrespective of crystallinity, possessing a super-
structure of oriented fibrillar domains (parallel to the applied
stress) alternating with solid moieties, exhibit known hard elastic
properties.

Interaction of Liquids with Equilibrium Stress

Our past results [4,5] have revealed the significant role of the
fibrillar domain in these materials, which we feel involves a surface
energy change within the fibrils. Such material should be sensitive
to changes in environmental surface tension.

A revealing experiment is the reaction of the hard elastic material, under stress, to liquid environments. When extended to a fixed
elongation, the stress relaxes to its equilibrium level which has been
considered as the "true" retractive force [4,5,11]. The reaction of
this stress to liquid environments is illustrated in Fig. 3 where
various types of interaction are shown. High surface tension liquids
such as water do not affect the stress level. The effect of a low
surface tension liquid, on the other hand, is exemplified by Freon-3
fluid. When the liquid is added, the equilibrium stress exhibits a
depression ($\Delta\sigma$), reaching a lower time-independent level. After re-
moving the liquid, a stress rise is observed and a third equilibrium
level is established. The difference between the third and initial
equilibrium stress level is defined as the residual stress depression
($\Delta\sigma_r$), illustrated in the figure.

In a recent publication [5], it has been shown that the liquid-
induced stress depression ($\Delta\sigma$), and the residual stress depression
($\Delta\sigma_r$) exhibited a strong relationship to the surface tension of the
environmental liquid. In this paper, we critically examine the stress
depression phenomenon as a function of the surface tension and viscosity of the interacting liquid, and the imposed deformation. The

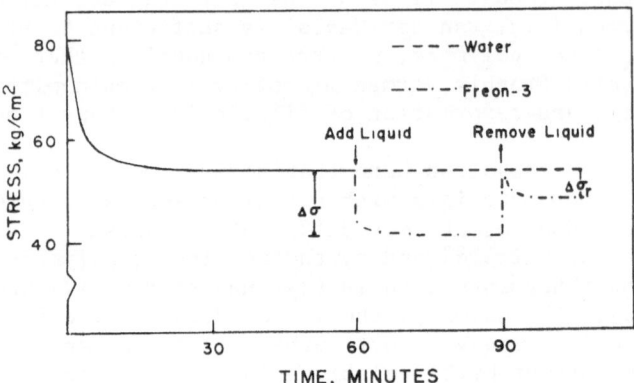

Figure 3: Stress-relaxation behavior of hard elastic HIPS in air for
one hour followed by immersion in water or Freon fluid
(arrow) for 30 minutes, then liquid removal (arrow).
Stress depréssion is clearly evident. Material strained
to 25%.

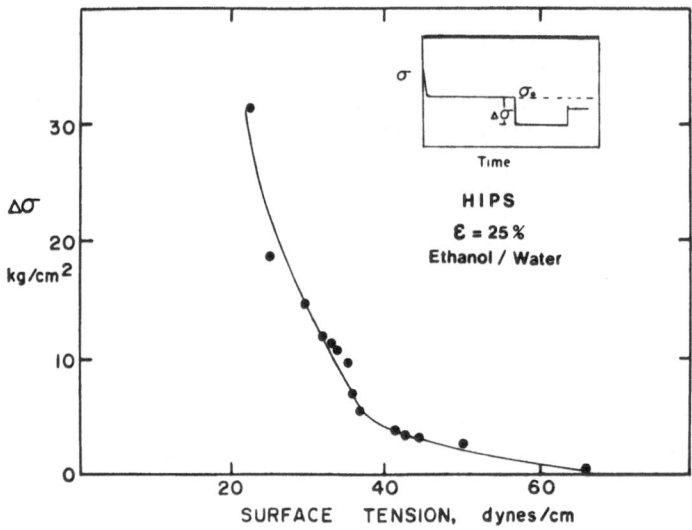

Figure 4: The effect of surface tension of the environmental liquid
(water-ethanol) on the stress elongation. Material
strained to 25%.

results are correlated with changes in the void volume fraction and the SAXS pattern as a function of elongation.

EXPERIMENTAL

Ordinary high-impact polystyrene, generously supplied by the Dow Chemical Company, was used as the material for all experiments. Dumbbell shaped specimens about 3 mm wide with a 16 mm gauge length were cut from compression molded sheets about 0.25 mm thick. The specimens were annealed in a vacuum oven at 86°C for 50 hours, then cooled at 10°C/hr. to room temperature to relieve internal residual pressures.

Hard elastic material was produced by straining the specimen in an Instron machine at a rate of 3×10^{-2} min^{-1} to an elongation of 40%, after which the specimen was unloaded and allowed to hang free for 10 minutes. The precrazed material was strained to the desired fixed elongation and the stress was allowed to relax for 1 hour. Samples were then immersed in liquid and the stress was allowed to stabilize for 30 minutes, after which the liquid was drained, resulting in a new stress level after a period of 10 minutes. Figure 3 illustrates the procedure.

Small angle x-ray measurements were performed using a Rigaku x-ray generator with rotating anode operated at 30 kv and 20mA. A vacuum camera with a 0.1 mm diameter pinhole collimator having a sample to film distance of 40 cm was used. Precrazed material was uniaxially strained in a stretcher to the desired elongation and clamped to a fixed position for the scattering experiment.

Void volume estimates were obtained by stretching preweighed samples to a fixed elongation and imbibing them with 5cS silicone oil. Excess oil was carefully wiped from the surfaces. The samples were then unloaded and oil released from the material was absorbed by preweighed pieces of filter paper. The total mass of the oil imbibed was then determined by weight difference.

RESULTS

To study the effects of the surface tension of liquids on the stress depression, $\Delta\sigma$, the precrazed material was strained to 25% elongation and the immersion procedure described above was followed using water-ethanol mixtures. The weight fraction of ethanol in the mixture was varied for different samples and the resultant stress depression was measured as a function of the surface tension of the mixture [12]. Figure 4 shows the relationship established between the surface tension, γ, of the immersing liquid and the stress depression, $\Delta\sigma$, exhibited by hard elastic material. The stress depression induced by pure ethanol (surface tension of 22.5 dyne/cm)

was rather large (31.3 Kg/cm^2). A dramatic decrease in $\Delta\sigma$ was observed when the surface tension of the immersing liquid was increased from
22.5 to 25.0 dynes/cm. This was followed by a more gradual, yet sharp,
decrease of $\Delta\sigma$ with increasing γ. The most interesting feature to
note is the sharp inflection point at a surface tension value of about
35-40 dynes/cm. This corresponds closely to the critical surface tension of wetting of polystyrene reported elsewhere [13]. Beyond this
transition, there was little change in $\Delta\sigma$ as γ increased. Note the
average change in slope at the inflection point from 10 to 0.25 Kg/cm.
dyne with increasing surface tension. When water (γ=72 dyne/cm) was
applied as the immersing liquid, no stress depression was observed.
Experiments using higher surface tension liquids including several
aqueous salt solutions, and mercury induced no change in the equilibrium stress level.

The phenomenon of liquid-induced depression of the equilibrium
stress was further studied as a function of deformation. The examination was undertaken by measuring $\Delta\sigma$ in ethanol at various fixed elongations. These values were normalized with the initial equilibrium
stress, σ_0, and the results are displayed in Figure 5. Note that an
essentially linear relationship between $\Delta\sigma/\sigma_0$ and elongation existed
for strain values up to 25%, after which the normalized stress depression remained constant.

The effect of fixed elongation on the ethanol-induced residual
stress depression is shown in Fig. 6. Note that the relationship is

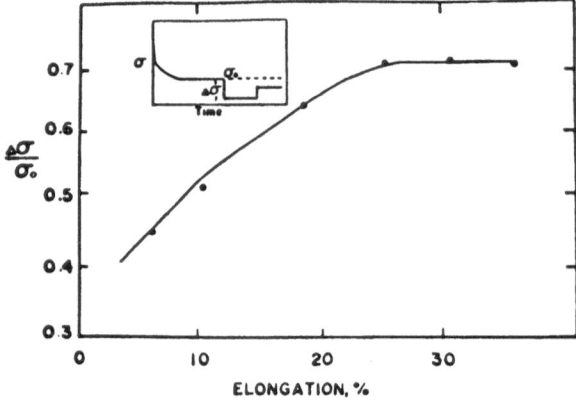

Figure 5: The relationship between the ethanol induced normalized
           stress depression and fixed elongation.

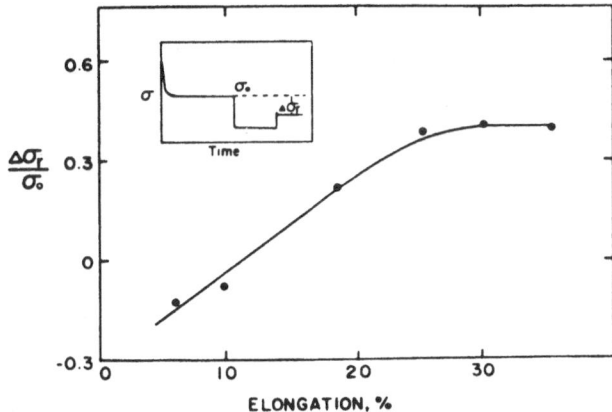

Figure 6:   The effect of fixed elongation of hard elastic HIPS on
            the normalized residual stress depression induced by
            ethanol.

very similar to the $\Delta\sigma/\sigma_0$ – elongation behavior. Below 15% elongation
however, upon removal of the ethanol environment, the stress rose to
a level exceeding the initial equilibrium stress. This is shown as a
negative residual stress depression.

    Figure 7 shows the stress depression behavior of hard elastic HIPS
immersed in silicone oils of different viscosities. Due to its low
evaporation rate, silicone does not discharge from crazed material
very easily after absorption. Once the lower stress level has been
reached, no stress rise was observed even after draining the oil and
exposing the imbibed material to air for over 24 hours. This imbibing
quality led to the use of silicone oil in determining the void volume
fraction in the crazed material as a function of strain. Void volume
estimates were determined by calculating the volume of oil imbibed
in the material, using the weight of the oil absorbed and its density
(0.93 g/cm^3), relative to the sample volume. Results of the void
volume fraction estimated at various initial elongations are plotted
in Fig. 8. At small elongations, the volume fraction increased at a
very slow rate up to about 15%. At this elongation, a rapid increase
was observed but did not appear to reach a plateau until 25% elongation. Although the volume fraction of the material appears to have
expanded more than twice its initial value within this range of elongation, no change in the cross-sectional area of the specimen was
observed.

    Small angle x-ray scattering patterns were obtained as a function
of constant elongation as shown in Fig. 9. At small elongations,
the SAX pattern showed a slight streaking along the meridian and no
other significant scattering. As the elongation was increased to 10%,
equatorial streaking appeared accompanied by the scattering on the

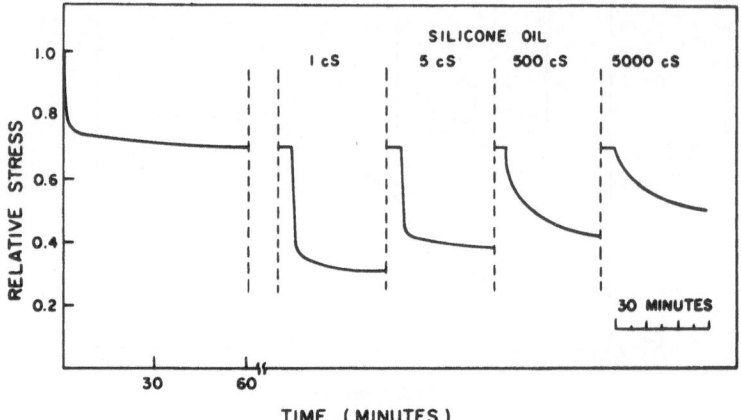

Figure 7:   The effect of the viscosity of silicone oil on the unre-
            laxed stress of precrazed HIPS at 25% fixed elongation.

meridian. The streaking increased at higher elongations becoming
well defined at about 20% elongation, reaching a maximum intensity
at the beginning of the stress plateau (25% elongation) and continued
at 30% elongation well within the stress plateau, although it
appeared more diffuse.

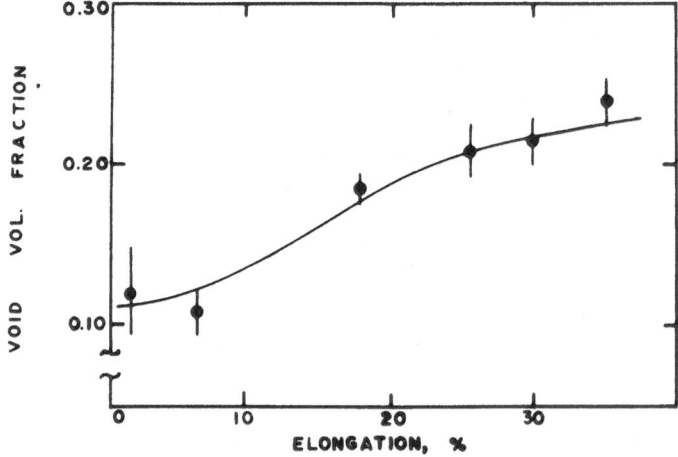

Figure 8.   The effect of fixed elongation on the void volume fraction
            of precrazed HIPS as estimated from adsorption measure-
            ments.

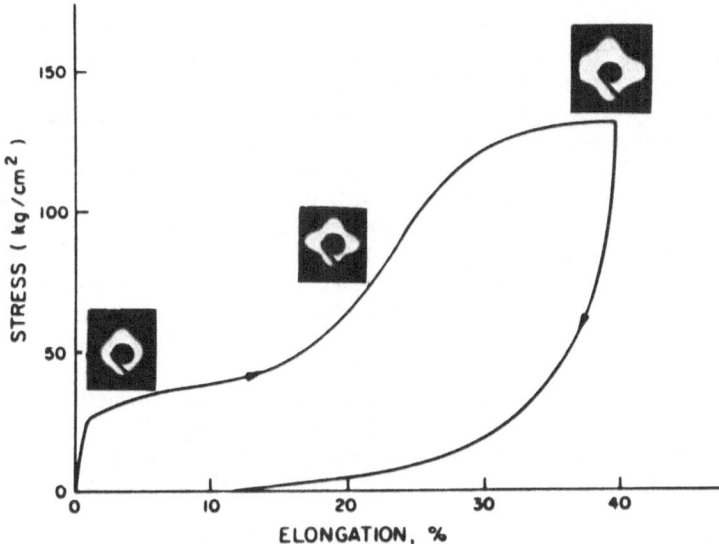

Figure 9:   Small-angle x-ray scattering patterns of precrazed HIPS
at various elongations.

DISCUSSION

This investigation was concerned with the elucidation of the
structural changes taking place during the deformation cycle of the
hard elastic material obtained from profusely precrazed HIPS. Of
course, the·interaction between the strained hard-elastic material
and the environmental fluid is a complex phenomenon; however, the
results clearly reveal several important features of the system.
The strong dependence of the equilibrium stress on the surface tension
of the environment lends support to the idea that the system functions
through an energy driven mechanism. The transition observed in the
$\Delta\sigma$-γ relationship at the critical wetting surface tension is a crucial factor in the process. The similarity between the stress-strain
relationship (Fig. 2) on one hand, and the stress depression-elonga-
tion relationship (Fig. 5) on the other, shows that the magnitude of
this interfacial energy closely associates with stress changes during
deformation. Hence, a possible increase of the fibrillar surface concentration is occurring during deformation. A sequential subfibrillation mechanism appears as a reasonable source for such an interfacial energy increase; however, craze boundary growth may not be
excluded. The latter is believed dominant at the stress plateau.

It may be contended that the described stress-depression behavior
results from the plasticization (or swelling) of the fibrillar component by the environmental liquid rather than surface energy effects.
The distinction between the two effects in the presence of an applied

stress has always been difficult. It is more so in our case where
the liquid-polymer interaction occurs on a scale very close to the
molecular level. That is, a solvent molecule acts upon a microfibril
with a diameter less than 100Å [14]. Nevertheless, the results reported here may provide insight into the nature of this complex problem. As pointed out earlier [5], fluids such as formamide, Freon-3
and silicone oil, which were found not to plasticize or swell polystyrene [15], do cause a significant stress depression. Further,
the stress rise occurring when the fluid is removed (Fig. 2) contra-
dicts our conventional understanding of plasticization or swelling.
The presence of a measurable residual stress depression upon liquid
removal may be associated with liquid entrapment due to the tortuous
nature of the fibrillar domain. Nevertheless, the negative residual
stress depression shown in Fig. 5 may rule out a plasticization argument, at least under the condition of this experiment. While this work
was in preparation, Brown [16] reported that similar effects were
observed during a study of the effect of surface tension of methanol-
water mixtures on the stress of environmental polystyrene crazes.

In contrast to conventional elastomers, when hard elastic material is stretched, there is very little or no decrease in cross-
sectional area. This behavior has been associated with the formation
of large interconnected voids, the volume of which was found to increase with elongation [1,2]. The entrapment of silicone oil by hard
elastic HIPS during the stress relaxation experiment (Fig. 7) prompted
its use in void volume determination. The void volume fraction estimated as a function of elongation (Fig. 8) behaves in an analogous
manner to the stress-strain relationship. Hard elastic polypropylene
was found to exhibit similar behavior [2]. The similarity between
the stress (Fig. 2), stress depression (Fig. 5), residual stress
depression (Fig. 6), and void volume fraction as a function of
elongation, knits together important features of the microdeformation-
al process(es) involved.

A qualitative analysis of the SAX patterns reported may shed some
light on the structural changes responsible for the hard elastic deformation of HIPS. The increase of the continuous equatorial streaking is due to a combination of increasing void content and inter-
fibrillar scattering. Very similar SAX patterns have been reported
for crazed atactic polystyrene [14], crazed polycarbonate [17], and
polypropylene which had been cold drawn and annealed with ends fixed,
then strained to 25% [18]. The resulting similarity of superstruc-
ture among these diverse materials as shown by SAXS lends additional
support to our belief that microfibrils play the dominant role in
"hard elastic" behavior.

CONCLUSIONS

Environmental liquid surface tension has a strong effect on the stress behavior of hard-elastic high-impact polystyrene. This effect is greatly manifested at surface tensions below the critical surface energy of the polymer and is strongly dependent on the structural changes induced during deformation. Critical examination of the results provides supportive evidence to the idea that the retractive force in hard elastic material stems from a reversible internal energy increase, possibly due to a sequential fibrillation mechanism. A simple method, using silicone oil absorption has been developed to estimate void volume fractions in crazed material.

ACKNOWLEDGMENTS

The authors wish to acknowledge generous financial support of the Office of Naval Research through Grant number N000-C-0795 and the National Science Foundation through Grant number DMR 77-24952. We are also indebted to Professor A. Chudnovsky for many useful discussions.

REFERENCES

1. Celanese Corporation of America, "Filamentous Material and a Process for its Manufacture", Belgium Pat. 650, 890 (January, 1965).

2. B. S. Sprague, "Relationship of Structure and Morphology to Properties of Hard Elastic Fibers and Films", in "The Solid State of Polymers", edited by P. H. Geil, E. Baer, Y. Wada, Marcel Dekker, Inc., New York (1974).

3. S. L. Cannon, G. B. McKenna and W. O. Statton, Macromol. Rev., 11, 209 (1976).

4. M. J. Miles and E. Baer, J. Mater. Sci., 14, 1254 (1979).

5. A. Moet, I. Palley and E. Baer, J. Appl. Phys., 51, 5175 (1980).

6. M. E. Mackay, T. G. Teng and J. M. Schultz, J. Mater. Sci., 14, 221 (1979).

7. P. Beahan, M. Bevis and D. Hull, Proc. R. Soc. Lond. A 343, 525 (1975).

8. L. Peters and H. J. Woods, in "The Mechanical Properties of Textile Fibers", ed. by R. Meredith, Interscience Publishers, Inc., N.Y. (1956).

9. T. Inoue, M. Moritani, T. Hashimoto and H. Kawai, Macromolecules, 4, 500 (1971).

10. J. Diamant and M. Shen, Polymer Preprints, 20(1), 250, ACS, April (1979).

11. M. J. Miles, J. Petermann and H. Gleiter, J. Macromol. Sci.-Phys., B12(4), 523 (1976).

12. A. W. Adamson, Physical Chemistry of Surfaces, Interscience Publishers, Inc., N.Y., p. 75 (1967).

13. J. Brandrup and E. H. Immergat, ed., Polymer Handbook, III-221 (1975).

14. H. R. Brown and E. J. Kramer, "Craze Microstructure from Small Angle X-ray Scattering (SAXS)", Report #4253, Dept. Mater. Sci. and Engineering, Cornell University (1980).

15. R. P. Kambour, C. L. Gruner and E. E. Romagosa, J. Polym. Sci.-Phys., 11, 1879 (1973).

16. H. R. Brown, private communication.

17. E. Paredes and E. W. Fisher, Makromol. Chemie, 180, 2707 (1979).

18. S. Natou and K. Azauma, J. Macromol. Sci.-Phys., B16(3), 435 (1979).

TRANSMISSION ELECTRON MICROSCOPE STUDIES OF POLYMERS

STAINED WITH RUTHENIUM AND OSMIUM TETROXIDE

J. S. Trent, P. R. Couchman, and J. I. Scheinbeim

Department of Mechanics and Material Science
Rutgers, The State University of New Jersey
Piscataway, New Jersey 08854

ABSTRACT

We demonstrate the effectiveness of ruthenium tetroxide as a
stain to enhance image contrast for transmission electron microscope studies of polymer morphology. The considerable microstructural detail that can be brought out by the stain is exemplified
by electron micrographs of thin films of incompatible blends of polystyrene with poly(methylmethacrylate). Crazes in strained solution-
cast thin films of these polymer blends and in a strained solvent-
cast thin film of high impact polystyrene can be seen readily, as
ruthenium tetroxide stains the polystyrene component preferentially.
Further, the stain darkens high-stress areas in these thin films.
The use of osmium tetroxide and ruthenium tetroxide to stain a
rubber-modified polymer of acrylonitrile and styrene reveals what
appear to be three phases rather than the two conventionally
accepted to occur. Morphological details of the spherulitic structure in Nylon 11 are made strikingly clear by the stain.

1.0 INTRODUCTION

Although the transmission electron microscope (TEM) has added
significantly to our knowledge of microstructural detail in polymer
systems, a number of difficulties arise when their thin films are
examined in this instrument. Foremost among these problems are inadequate morphological contrast and beam-induced structural changes,
including film damage. The occurrence of specimen structure changes
and damage can be controlled to some extent by operating the microscope at low intensities and by keeping the specimen cold during
examination. Image contrast can be enhanced by the use of selective

205

stains with a high electron density relative to the specimen material.

Kato (1,2) was the first to overcome the image contrast problem by the preferential staining of polymer using osmium tetroxide (OsO_4). Unfortunately, the use of OsO_4 is limited to those polymers having some level of unsaturation and, until recently, there were no known and convenient selective strains for polystyrene (PS) and other types of saturated polymers. Recently, however, there have been two independent discoveries (3-5) that ruthenium tetroxide (RuO_4) is an effective staining agent for the examination of both saturated and unsaturated polymeric systems in the TEM. Vitali and Montani (3) observed improved image contrast for polybutadiene latices and microtomed specimens of a butadiene-modified polymer of acrylonitrile and styrene (ABS), and an acrylonitrile-styrene-acrylonitrile (ASA) polymer. Separately, we have reported on the efficacy of RuO_4 vapor-staining for TEM studies of various solvent-cast polymer films, in particular polystyrene and its blends (4,5). We observed that RuO_4 is an effective stain for polymers containing ether, alcohol, aromatic or amine moieties in their unit structure, and was unreactive towards poly(methylmethacrylate) (PMMA), poly-(vinyl chloride) (PVC), poly(vinylidene fluoride) (PVF_2) and poly-acrylonitrile (PAN) (5). High density polyethylene (HDPE) and both isotactic and atactic polyprolylene (PP) were lightly stained by RuO_4. Strikingly solvent-dependent morphological changes were shown to occur for solution-cast films of 10%PS/90%PMMA.

The present work is a review of our previous TEM observations (4,5) and an initial report on new inquiries into the morphology of solvent-cast films of ABS and Nylon 11.

2.0 EXPERIMENTAL PROCEDURE

Clean glass slides (50mm x 10mm) were dipped into a 1-2% solu-tion (by weight) of polymer in solvent; specimens were vapor-stained after evaporation of the solvent. A detailed discussion of the solvent evaporation procedure and the staining technique is given elsewhere (1,4,5). A concentration of 1% by weight of polymer in solvent yielded films of thickness between 3000-5000Å. A Hitachi HU11-A electron microscope was used for this study, at an accelerating voltage of 75 Kv.

The ABS polymer for the TEM study (obtained from the Dow Chem-ical Co.) was synthesized at the azotropic composition (25% acrylo-nitrile and 75% styrene, by weight). The rubber content was 13 wt%, with an average particle size of 0.5 µm. Thin films of high impact polystyrene (HIPS), ABS, and Nylon 11 were cast from 1% solutions (by weight) of polymer in, respectively, toluene, ethyl acetate, and a 50/50 mixture of phenol/formic acid. Toluene, N,N-dimethyl-

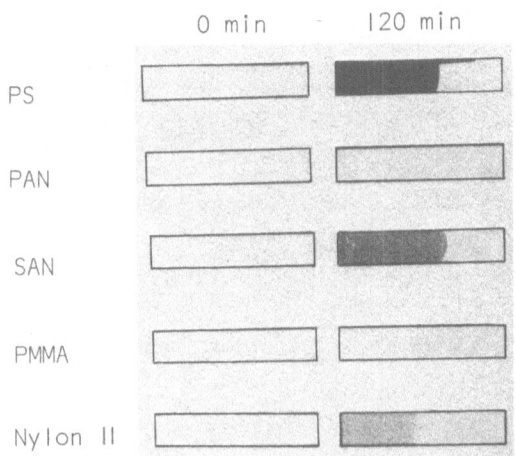

Figure 1.   Optical photographs of unstained PS, PAN, SAN, PMMA, and
            Nylon 11 and these polymers stained with RuO_4 for 120
            minutes. The films were stained on glass slides which
            are outlined in black.

acetamide and N,N-dimethylformamide were used as solvents for in turn
PS and PMMA, PAN, and a styrene-acrylonitrile polymer (SAN).

3.0   RESULTS AND DISCUSSION

     Figure 1 shows a comparison of unstained films of PS, PAN, SAN,
PMMA and Nylon 11, and those stained for 120 minutes with RuO_4 vapor.
It is apparent that PS and SAN are very sensitive to RuO_4 vapor.
Nylon 11, while less sensitive, was darkened moderately. Films
made of PAN and PMMA appear unaffected by RuO_4 vapor. The selective
action of RuO_4 provides a method for direct and convenient observation of polymer microstructure for those polymer blends in which
one component is stained preferentially.

     Thin films of a PS/PMMA blend of ratios 90/10 and 10/90 (by
weight) were stained and examined in the TEM (Figure 2). Both films
were vapor-stained with RuO_4 for 30 minutes. Figure 2(a) shows the
morphology of a 90%PS/10%PMMA blend where the dark regions are the
PS-rich phase. In the light PMMA regions it is evident that there
are small precipitates of PS. Figure 2(b) shows the morphology of

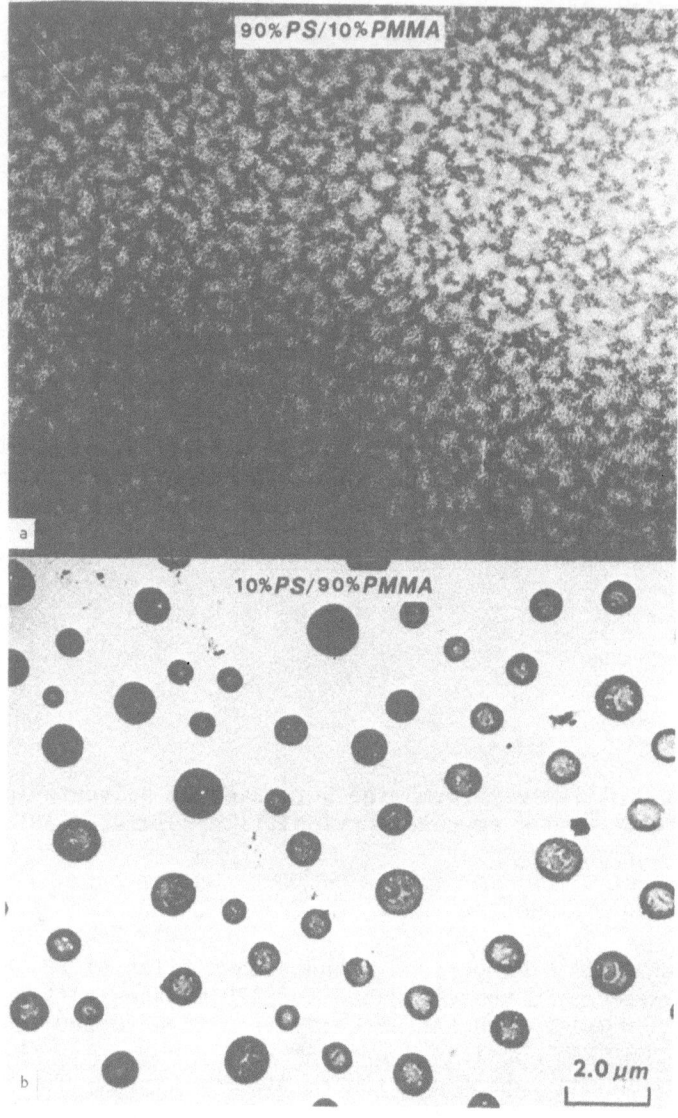

Figure 2. Electron micrograph of (a) 90% PS/10% PMMA and (b) 10%PS/
           90% PMMA film stained for 30 minutes with RuO_4. PS is
           the darker phase.

a 10%PS/90%PMMA blend where again the dark regions are PS-rich.
Here, droplets of PS are present in the PMMA matrix. The internal
structure of these droplets appear in turn to be a composite of PS
and PMMA. As is the case with HIPS, the inclusions (here of PMMA)
within the (PS) droplets raises the volume fraction of droplets
present in the (PMMA) matrix. The reverse is true for the PMMA
phase in the PS matrix, as seen in Figure 2(a). The fine structural
detail brought out by vapor staining with RuO_4 is clearly demonstrated
by the various micrographs.

     To examine crazes in blends, film-covered copper grids were
strained with a pair of tweezers and then vapor-stained with either
RuO_4 or OsO_4. Figure 3 is an electron micrograph of a craze which
has developed in a stressed film of a 4%PS/96%PMMA blend. In this
film, stained for 15 minutes with RuO_4 vapor, the craze fibrils
(ranging in diameter from 100-300Å) are seen clearly, as is their
branching. An added feature of the stain is that it tends to
harden the film and fibrils, as does osmium tetroxide with unsaturated polymer (1,2), thereby improving film stability during exposure
to the electron beam.

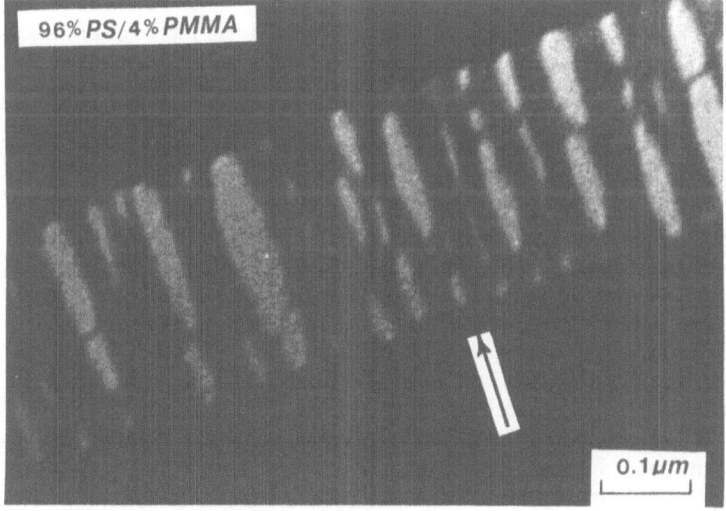

Figure 3. Electron micrograph of a craze formed in a 4%PS/96%PMMA
           film stained for 15 minutes with RuO_4. Arrow indicates
           direction of applied stress.

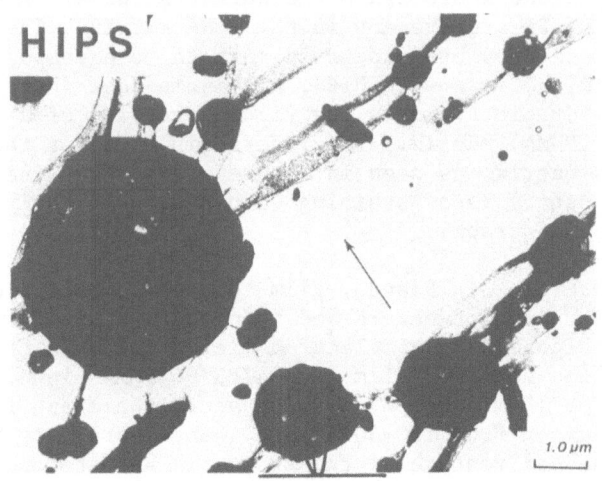

Figure 4.   Electron micrograph of a stressed film of HIPS stained
            for 45 seconds with OsO$_4$. Arrow indicates direction of
            applied stress.

     Figure 4 is an electron micrograph of a stressed HIPS film
stained for approximately 45 seconds with OsO$_4$ vapor. Since OsO$_4$
reacts with the double bonds in the polybutadiene phase and not
with the aromatic ring in the PS matrix only the rubber is darkened.
Occlusions of PS can be seen within the rubber particles along with
crazes which have been initiated at the PS-rubber interface and
propagated perpendicular to the direction of applied stress. The
crazes have formed what appears to be a cell structure. When a
stressed film of HIPS is stained with RuO$_4$ vapor the detail observed
is quite different, as seen in Figure 5.

     Figure 5 is an electron micrograph of a stressed HIPS film
showing a craze which has interacted with a rubber particle. After
straining, this film was stained for 5 minutes with RuO$_4$ vapor.
Staining with this oxide permits detection of fibrils within the
craze even down to the craze tip (denoted by arrow a) as RuO$_4$ stains
both the PS matrix and the rubber particles. The rubber particles
are not stained to the extent that would occur with OsO$_4$ (Figure 4),
permitting observation of their interaction with the craze. As the
craze propagated through the circular rubber particle (denoted by
arrow b), the rubber separated from the PS matrix (absorbing energy)
causing a split in the craze. When the craze interacts with the
next rubber particle this split is no longer present.

     As we discovered, the combined use of both OsO$_4$ and RuO$_4$ as
staining agents can be advantageous. Figure 6 is an electron micrograph of an ABS film first stained with OsO$_4$ vapor for 1 minute and

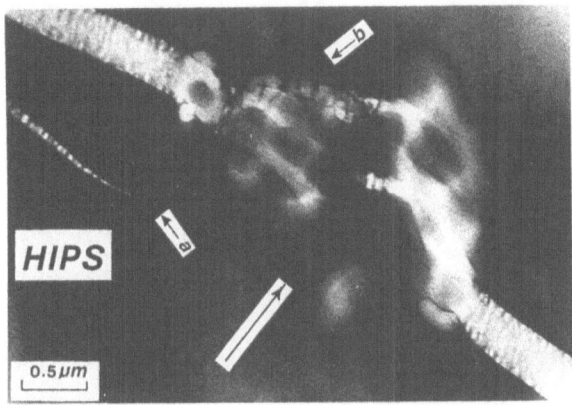

Figure 5.   Electron micrograph of a craze which has propagated
            through several rubber particles in HIPS film stained
            for 5 minutes with RuO_4.   Arrow indicates direction of
            applied stress.

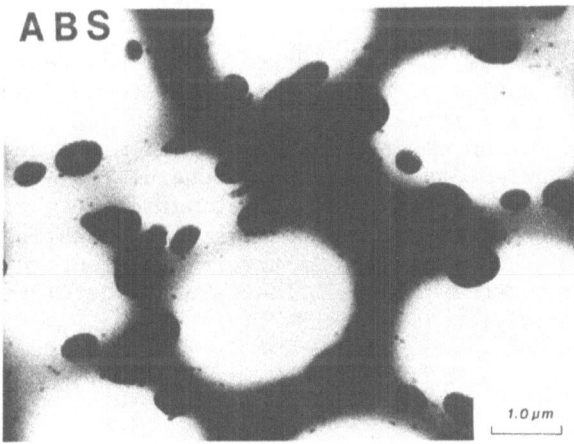

Figure 6.   Electron micrograph of ABS film stained first with OsO_4
            for 1 minute and then with RuO_4 for 30 minutes.

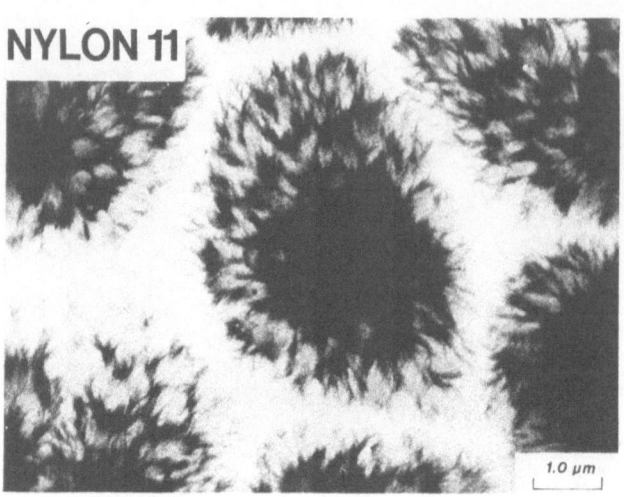

Figure 7.   Electron micrograph of a Nylon 11 film stained for 30
            minutes with RuO_4.

then with RuO_4 vapor for 30 minutes. Surprisingly, the stain appears to reveal three types of regions instead of the conventionally
recognized two-phase system for ABS. This film was heat-treated
at $95°C$ for approximately 20 hours before staining. Since OsO_4
stains only the rubber phase (1,2) and RuO_4 stains both rubber and
PS but not PAN (Figures 1 and 5) we suggest that the darkest particles in Figure 6 are the rubber phase, the moderately dark areas
may be a PS-rich phase and the light areas a PAN-rich phase. We
know of no other observations of three-phase morphology in ABS, a
clear understanding of which is not presently available.

     When Nylon 11 was found to darken upon exposure to RuO_4 vapor
(Figure 1), a thin film of this polymer was stained and examined
in the TEM. Figure 7 is an electron micrograph of Nylon 11 stained
with RuO_4 vapor for 30 minutes. The nylon film was heat-treated
at $50°C$ for 24 hours before staining. Spherulites approximately
5 μm in diameter can be seen. It is not yet clear what moieties
the RuO_4 interacts with, but our observations established that the
stain stabilized the Nylon 11 film toward the electron beam. It
is now possible to study thermal effects and spherulitic growth
in Nylon 11 and polymers such as HDPE and PP.

CONCLUSIONS

     It has been established that a staining procedure using RuO_4
can improve image contrast of both saturated and unsaturated poly-

mers, including PS, SAN, HIPS, ABS and Nylon 11. In addition, OsO_4 and RuO_4 used together can improve image contrast further, as seen for ABS.[4]

ACKNOWLEDGMENT

The authors wish to thank the National Science Foundation Polymers Program, for their generous financial support of this work under Grant DMR 79-15175.

REFERENCES

1. K. Kato, J. Electron Microscopy, 14, 220 (1965).
2. K. Kato, Polym. Eng. Sci., 7, 38 (1967).
3. R. Vitali and E. Montani, Polymer 21, 1220 (1980).
4. J.S. Trent, J.I. Scheinbeim and P.R. Couchman, J. Polym. Sci., Polym. Lett. Edn., 19, 315 (1981).
5. J.S. Trent, J.I. Scheinbeim and P.R. Couchman, submitted for publication.

THE TIME-TEMPERATURE-TRANSFORMATION (TTT) STATE DIAGRAM

AND ITS ROLE IN DETERMINING STRUCTURE/PROPERTY RELATIONSHIPS

IN THERMOSETTING SYSTEMS

John K. Gillham

Polymer Materials Program
Department of Chemical Engineering
Princeton University
Princeton, New Jersey 08544

ABSTRACT

A generalized time-temperature-transformation (TTT) state diagram for the thermosetting process is presented in which the four physical states encountered (i.e., liquid, rubber, ungelled glass and gelled glass) are related to the time and temperature of cure. Gelation and vitrification, as a consequence of quenching morphological development and chemical conversion respectively, are discussed with respect to control of material properties.

INTRODUCTION

Composite materials involving an organic matrix reinforced with continuous filaments having a high tensile modulus and strength are important in applications requiring light-strong structures. In such materials the organic matrix is generally formed by the chemical conversion of a reactive fluid to a solid in the thermosetting process. Although thermoplastic materials can also be used, they are of limited application because of the high viscosity of their melts, their relative dimensional instability under load, and their unsuitable composite performance above the load-limiting transitions of the organic matrix (i.e., the glass transition temperature, T_g, for amorphous and the melting temperature, T_m, for semicrystalline polymers).

The most important thermosetting matrices involve network systems such as the epoxies and semi-ladder polymers such as the

215

polyimides. The proper exploitation of these materials is currently
restricted because of the unsatisfactory state of the scientific and
technical information available concerning the interdependence of
their molecular structure and their morphological and mechanical
properties. The fundamental reasons are a lack of understanding of
the cure process and of the non-equilibrium nature of the glassy
state. However, from the experimental point of view, they are also
inherently difficult materials to study. They are infusible and
insoluble and are therefore synthesized and fabricated in one operation: because of this, their chemistry and physics are strongly
coupled. The amorphous nature of the materials also restricts the
applicability of diffraction and morphological techniques that can be
used with crystalline and oriented samples.

The very intractability which makes the characterization of
thermosetting materials difficult is associated with the reasons for
their superior engineering behavior. A material property of particular importance that is related to the nature of the molecular networks is their dimensional stability under mechanical stress.
However, in the unreinforced state the materials are brittle and they
must therefore be used in structural applications in the form of
fiber-reinforced composites or chemically produced two-phase
materials. The current interest in composites makes it essential to
understand the physical properties of these organic matrices in
relation to their chemistry. Again, from the practical point of
view, it is to be noted that homogeneous unreinforced specimens are
often difficult to prepare in a defect-free state for testing because
of residual curing and thermal shrinkage stresses, bubble inclusions
introduced during cure, and surface defects introduced during test
specimen preparation. In addition, the chemical approach to the
study of molecular structure-bulk property relations has been made
difficult because of the ubiquitous use of impure reactants, pro-
prietary formulations and arbitrary curing conditions. Each of these
factors becomes of greater importance as the performance expected
from the composite is increased.

Even with pure reactants the complexity and competing nature of
the chemical reactions involved in synthesizing the network materials
would make molecular structure-bulk property correlations difficult
to obtain. What is required is more general understanding of the key
relationships between the process of cure and the properties of the
cured state. It is to this point that the present article is
directed.

Recent research [1-5] has indicated that a Time-Temperature-
Transformation diagram (analogous to the TTT diagrams that have been
employed for many years in metallurgical processing) may be used to
provide an intellectual framework within which an understanding of
the physical properties of thermosetting matrices may be achieved.
In the technical discussion presented below, the significance of this

diagram is discussed and it is used to explain practices and
phenomena prevalent in the technology of thermosets.

TIME-TEMPERATURE-TRANSFORMATION (TTT) STATE DIAGRAMS

Time-Temperature-Transformation (TTT) diagrams have played an
important role in the control of the properties of metals by permitting thermal history paths to be chosen so that a desired microstructure can be obtained. The diagrams are specific to a particular
material composition and considerable insight into the design of
alloys can be achieved once the effects of additions of alloying
elements on the TTT diagram have been explored. Since thermosetting
polymeric systems are prepared in situ, the availability of an equivalent diagram for either the pure matrix material, or a matrix containing impurities such as a dispersed rubber phase, would be of
considerable technological importance. Such a diagram would permit
time-temperature paths for cure to be chosen so that gelation, vitrification and phase separation occurred in a controlled manner and
consequently give rise to predictable properties of the thermosetting
matrix.

Gelation and vitrification are two macroscopic phenomena which
are encountered as a consequence of chemical reactions which convert
a fluid to a solid in the thermosetting process. On the molecular
level, gelation corresponds to the incipient formation of branched
molecules of very high molecular weight. Macroscopically this process is accompanied by a dramatic increase in viscosity and a corresponding decrease in the condensed phase diffusional processes and in
material processibility. In principle, gelation starts at a fixed
chemical conversion that can be predicted from the functionality of
the reactants. Eventually the total mass of material can be regarded
as one molecule. This network structure will be an elastomer at a
given temperature if the segments between junction points of the
network are flexible. If these segments are immobilized by further
chemical reaction, or by cooling, the structure will change to the
glassy (vitrified) state.

Vitrification, which usually follows gelation, occurs as a consequence of increasing molecular weight and further crosslinking
causing a reduction in the degrees of freedom of the network.
Vitrification can further retard (or quench) chemical reactions in
the matrix.

The overall transformation from liquid to gel to rubber to glass
that occurs as a result of chemical reactions in the thermosetting
process is termed "cure". The properties of the final material are
intimately related to the details of the curing process. In particular they depend upon the interplay between such factors as the chemical reactants involved, their mutual solubility, their viscosity

prior to gelation, the volatility of the reactants and byproducts, gelation, phase separation, vitrification, overall chemical conversion, the details of the time-temperature path of the curing reaction, and the limits of the thermal stability of the materials involved.

Figure 1 [3] shows a generalized TTT diagram obtained from isothermal experiments for a typical thermosetting process that does not involve phase separation. It displays the four distinct material states (liquid, elastomer, ungelled glass and gelled glass) that are

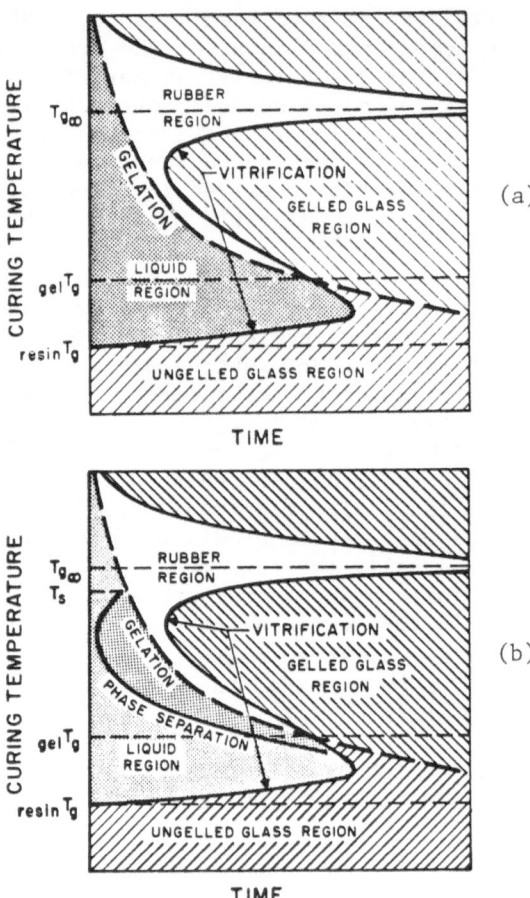

Fig. 1.   Schematic Time-Temperature-Transformation (TTT) diagrams obtained isothermally for the curing process. a) A thermosetting system not involve phase separation. b) A thermosetting system in which a rubber phase may separate during cure. (T_s is the maximum temperature for phase separation to occur prior to gelation.)

encountered during cure. Three critical temperatures are also displayed on the diagram. These are $T_{g\infty}$, the maximum glass transition temperature of the fully cured system, $_{gel}T_g$, the isothermal temperature at which gelation and vitrification occur simultaneously, and $_{resin}T_g$, the glass transition temperature of the reactants.

When a thermosetting material is cured isothermally above $T_{g\infty}$, the liquid gels to form an elastomer but it will not vitrify in the absence of degradation. (Vitrification due to degradation is shown in Fig. 1). An isothermal cure at an intermediate temperature between $_{gel}T_g$ and $T_{g\infty}$ will cause the material first to gel and then to vitrify. If chemical reactions are quenched by vitrification it follows for this case that the glass transition temperature will equal the temperature of cure and that such a material will not be fully cured. At temperatures below $_{gel}T_g$ but above $_{resin}T_g$ the viscous curing liquid can vitrify simply by an increase of molecular weight and, if chemical reactions are quenched by vitrification, the material need not gel.

It is immediately apparent that structure-property relationships will only be meaningful if the material is fully reacted. This is generally only possible by curing above $T_{g\infty}$.

As indicated in Fig. 1 the time to vitrify passes through a minimum between $_{gel}T_g$ and $T_{g\infty}$. This behavior reflects the competition between the increased rate constants for reaction and the increasing chemical conversion required to achieve vitrification as the temperature is increased.

The cure TTT diagram of Fig. 1a can be extended (see Fig. 1b) to include two phase systems such as a rubber incorporated in an inherently brittle polymeric material in order to increase the toughness of the composite. The curing of such rubber-modified systems may involve a change from an initially homogeneous solution to a heterogeneous multiphase morphology. Gelation may arrest the development of the rubber second phase and therefore procedures which alter the time and temperature to gelation can be used to control the material properties. Control of the time-temperature history of the material during cure is a method of achieving the desired degree of phase separation, but a knowledge of the TTT diagram is prerequisite for such a procedure. Since the nucleation and growth of the rubber phase involves a balance between nucleus formation and matter transport, the degree of phase separation achieved in an isothermal process would be expected to show a maximum at a temperature between that for which thermodynamics favors the solubility of the rubber in the matrix and the $_{resin}T_g$ of the matrix. Careful control of the cure temperature will also permit the size and number of particles per unit volume of the dispersed rubber phase to be modified and hence have a strong effect on the mechanical properties. Gelation can give rise to an abrupt decrease in the diffusion of rubbery

material to the growing second phase domains and hence effectively
quench the development of the second phase morphology. Evidence has
been presented [6,7] to show that improved material toughness arises
in rubber-modified systems in which part of the rubber is phase-
separated and part is trapped in solution in the matrix. The path
taken on the TTT diagram must therefore be chosen so as to balance
the distribution of the rubber between the two phases.

A Continuous Heating Transformation (CHT) State Diagram which is
analogous to the isothermally obtained TTT state diagram can be
obtained experimentally from a series of temperature scans at different rates from below the glass transition temperature of the reactants ($_{Resin}T_g$) to above $T_{g\infty}$. A typical scan for a homogeneous reactive system will reveal in sequence: relaxations in the glassy
state, $_{Resin}T_g$, gelation, vitrification and (in the presence of some
types of degradation) revitrification. After vitrification on cure,
in these scans, the glass transition temperature will in principle
equal the instantaneous scanning temperature until the rate of chemical reaction is not sufficient to overcome the increased segmental
mobility of the developing network, at which temperature the material
will devitrify.

The above discussion has been intended to introduce the concept
of TTT diagrams for thermosetting materials and to indicate their
utility in the control of processing that will influence the mechanical properties of these materials. To further illustrate this point
the discussion immediately below uses the TTT cure diagram to explain
a number of practices current in the field of thermosets. The final
section will include several specific examples of the effect of
undercure versus more fully developed cure on material behavior, and
the influence of gel time on morphology.

If the storage temperature is below $_{gel}T_g$ a reactive fluid
material will convert to a vitrified solid of low molecular weight
which is stable and can be later liquified by heat and processed.
Above $_{gel}T_g$ the stored material will have a finite shelf-life for
subsequent processing since gelation will occur before vitrification.
(A gelled material does not flow in the usual sense.) This concept
lies at the basis of a widespread technology which includes thermosetting molding compounds and "prepregs" with latent reactivity.

In general, if $T_{cure} < T_{g\infty}$, a reactive material will vitrify and
full chemical conversion will be prevented. The material will then
usually need to be post-cured above $T_{g\infty}$ for development of optimum
properties. For highly crosslinkable or rigid chain polymeric materials T_g can be above the limits of thermal stability in which case
full chemical conversion of the original network-forming reactions
would not usually be attainable. For composite materials in which a
component other than the cured resin is thermally sensitive, $T_{g\infty}$ for
the thermosetting resin should be below temperatures which would lead
to damage of any part of the assembly.

Time-Temperature-Transformation diagrams have been essential in the exploitation of metallic systems and particularly in the control of their mechanical properties in the alloyed state. It is highly desirable to be able to exert equivalent control over the properties of thermosetting polymeric systems. In order to obtain such control it will be necessary to develop TTT diagrams for specific systems using pure reactants and to extend these studies to include the addition of deliberate impurities, such as rubbers, to these materials. On the basis of these diagrams, thermal history paths for the cure process may be chosen and desired final morphologies achieved. The final step in this process is to relate the materials' morphologies to their mechanical properties such as ductile-brittle behavior, toughness and fatigue resistance.

TORSIONAL BRAID ANALYSIS (TBA) - A METHOD FOR CONSTRUCTING TTT AND CHT STATE DIAGRAMS

An automated, free-hanging, freely decaying torsion pendulum has been developed [1,8] which permits monitoring of the changes which occur throughout cure by using a support (e.g. braid) impregnated with the reactive system (TBA).

The TTT diagram (Fig. 1a) can be generated by measuring times to gel and to vitrify at a series of isothermal temperatures. These transformation times have been obtained using TBA from measurements of peaks in the mechanical damping as a function of time which correspond to points of inflection in the rigidity curves. Complementary CHT diagrams have been generated by scanning the temperature range from below the glass transition temperature of the reactant mixture, $_{resin}T_g$, to above $T_{g\infty}$ at a series of constant heating rates. The concept of TTT and CHT diagrams for thermosetting systems has been developed using TBA.

Details of procedures and results on structure-property relationships of thermosetting systems are provided in recent publications [1-12].

An advantage of a freely suspended torsion pendulum specimen over forced systems is its sensitivity which is the consequence of the free end. This has led to the revelation of fine details in the dynamic mechanical spectra which suggest that gelation, as measured by TBA, is a multistep process [4,10,11].

EFFECT OF EXTENT OF CURE ON MATERIAL BEHAVIOR

In principle, cure at a temperature T_{cure} which is below $T_{g\infty}$ will lead to the glass transition temperature $T_g = T_{cure}$. Post-cure above $T_{g\infty}$ will lead to $T_g = T_{g\infty}$. An epoxy was cured according to the manufacturer's specifications and yielded the thermomechanical TBA behavior "before post-cure" shown in Fig. 2 (which lists experimental details and a summary of transitions). It is apparent that the cure cycle was not sufficient to have $T_g = T_{cure}$. Post-cure resulted in significant increase in T_g as well as change in the viscoelastic behavior (e.g. the damping behavior) below the glass transition. In particular, a small but significant decrease in the rigidity (i.e., modulus if not dimensional changes occur on post-cure) occurred in the glassy state at, for example 0°C, in consequence of the post-cure. This is a result of the non-equilibrium nature of the glassy state. Cooling at the same rate through a glass transition leads to a supercooled liquid which is further from equilibrium at, e.g. 0°C, the higher the T_g.

Consider (Fig. 3) two specimens (1 and 2), one cured above $T_{g\infty}$ at temperature $_2T_{cure}$, the other cured below $T_{g\infty}$ at temperature $_1T_{cure}$. Specimen 1 vitrified on cure to give a glass transition equal to the temperature of cure. Specimen 2 reacted completely to give the maximum glass transition temperature ($T_{g\infty}$). In the absence

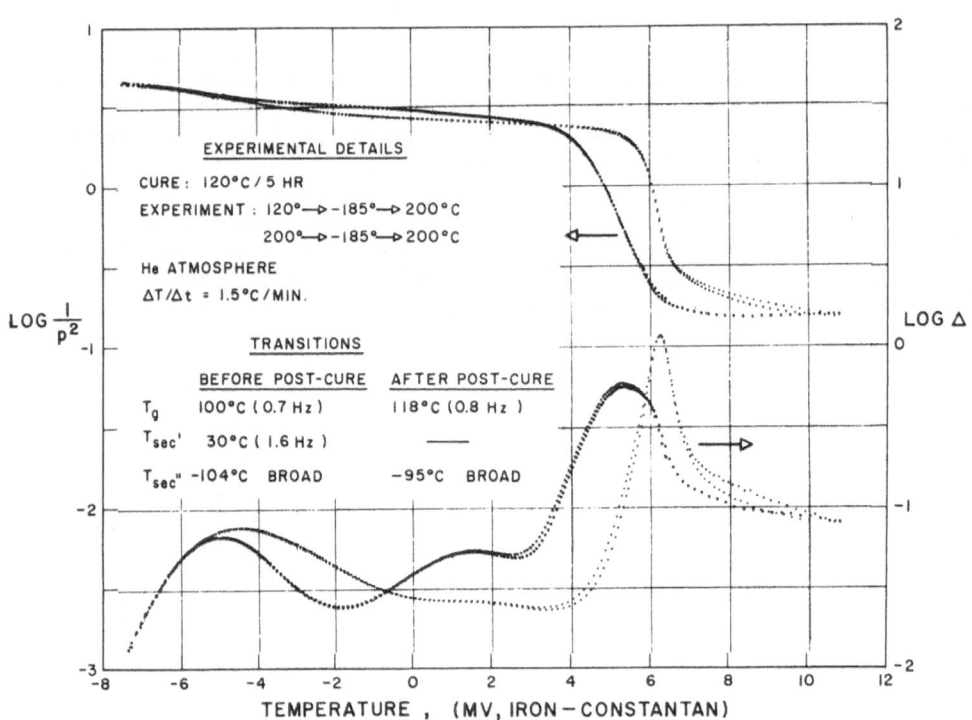

Fig. 2.   Thermomechanical behavior (TBA) of an epoxy after
          recommended cure and after post-cure. Note the decreased
          rigidity at 0°C (0mV) after post-cure.

of further reaction the specific volume of the specimen cured at the lower temperature will be higher at $_2T_{cure}$ than that cured at $_2T_{cure}$ (due to lower crosslink density and more segmental free ends). The diagram, Fig. 3 [13], shows that cooling of the more completely re-acted material (at equal rates) results in a higher T_g and indicates how a reacted material (at equal rates) results in a higher T_g and indicates how a higher specific volume results in the glassy state. The fundamental reason is that the material with the higher T_g is further from equilibrium at, e.g., room temperature (RT).

The higher specific volume at RT of the more highly crosslinked material is held responsible for its lower density and lower modulus at RT and greater water absorption on immersion at RT [5,13,14].

EFFECT OF GELATION ON MATERIAL PROPERTIES IN TWO–PHASE SYSTEMS

The curing of rubber-modified epoxy systems can involve change from an initially homogeneous solution of reactants containing epoxy resin, curing agent and reactive liquid rubber (e.g. carboxy-terminated acrylonitrile butadiene copolymer) to a two phase system having rubber particles dispersed in an epoxy matrix. The modified schematic TTT state diagram of Fig. 1b includes the locus of cloud point measurements which demarcates the onset of phase separation as determined visually. Growth of the rubbery domains is considered to continue until gelation. Different morphologies arise from cure at

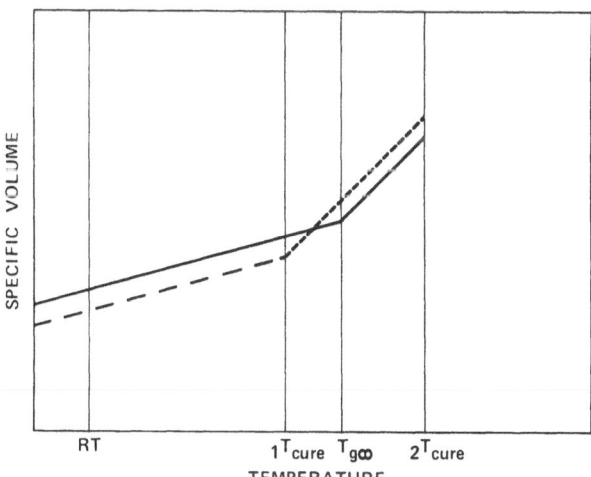

Fig. 3.   Schematic: Specific volume versus temperature in the absence of chemical reaction. Cured at temperature $_1T_{cure}$: dashed line. Cured at temperature $_2T_{cure}$: solid line. Note the higher specific volume at room temperature of the more highly reacted system.

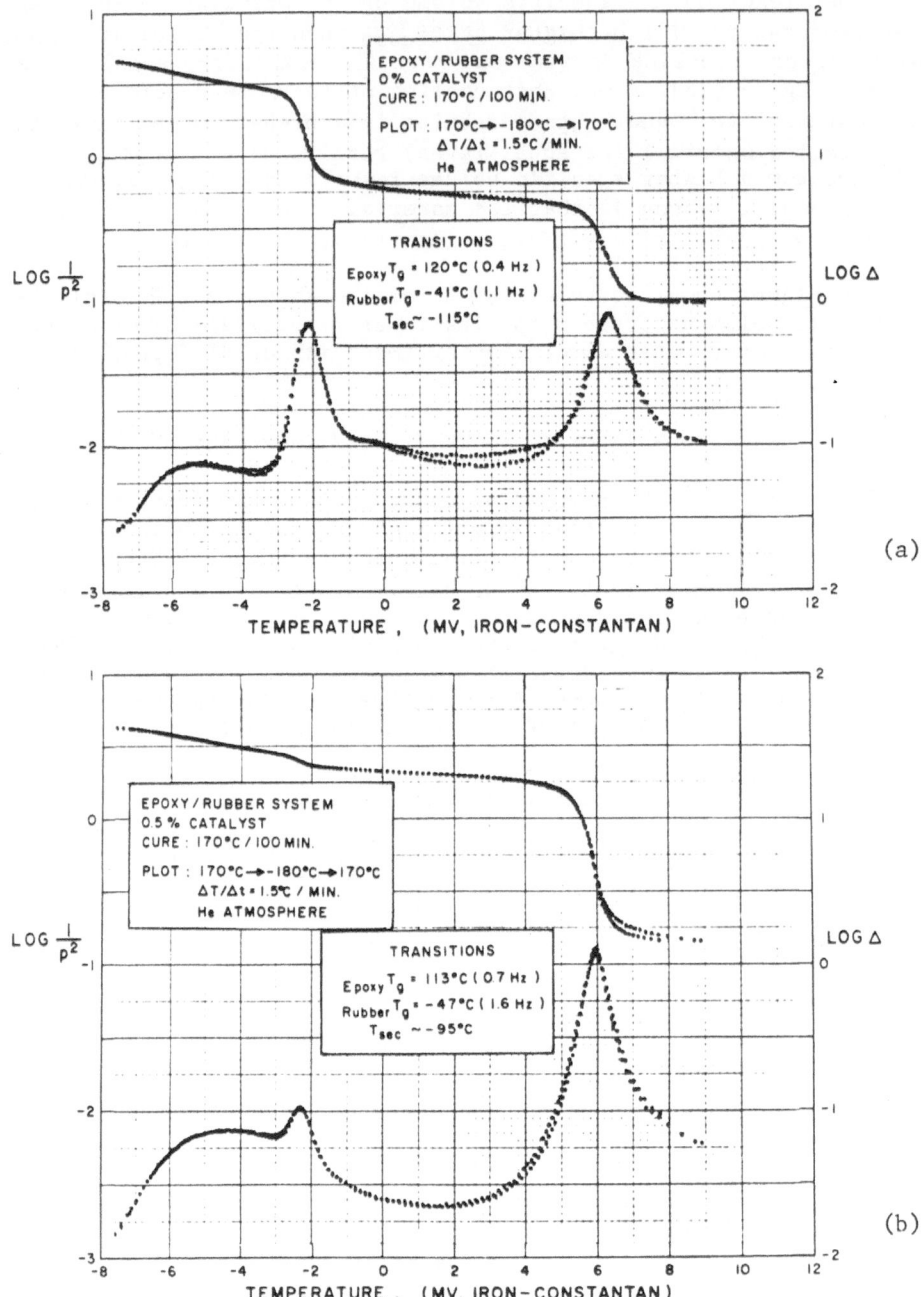

Fig. 4. Thermomechanical behavior (TBA) of an epoxy-rubber system
         following identical time-temperature cure paths. a) Zero
         parts per hundred of catalyst. b) 0.5 parts per hundred of
         catalyst.

different temperatures due to the influence of temperature on the competition of thermodynamic and kinetic factors. For example, cure above temperature T_S (Fig. 1b), the temperature above which phase separation does not occur prior to gelation, leads to an optically transparent material in which rubber is finely dispersed in the matrix. Cure at lower temperatures than T_S leads to visual phase separation, the number and size of the rubbery domains depending on the temperature. The morphology developed controls the material properties.

An example of the influence of gelation time on morphology is made evident by comparison of the thermomechanical TBA behavior of a rubber-modified epoxy cured without (Fig. 4a) and with (Fig. 4b) catalyst [12]. The glass transition temperature of the rubber is much more dominant in the sample cured without catalyst. This suggests that the extent of phase separation depends on the time available for phase separation which is limited by the process of gelation. The higher glass transition temperature of the epoxy for the sample cured with the longer gelation time also suggests more complete separation of the two phases.

REFERENCES

1. J. K. Gillham, Polym. Eng. Sci., 19 (1979) 676.
2. M. J. Doyle et al., Polym. Eng. Sci., 19 (1979) 687.
3. J. K. Gillham, SPE, Proceedings, ANTEC Meeting, N.Y. (1980) 268.
4. J. B. Enns et al., ACS, Prepr., Div. Org. Coat. Plast. Chem., 43 (1980) 669.
5. M. J. Doyle et al., ACS, Prepr., Div. Org. Coat. Plast. Chem., 43 (1980) 677.
6. L. T. Manzione et al., ACS, Prepr., Div. Org. Coat. Plast. Chem., 41 (1979) 364.
7. L. T. Manzione et al., ACS, Prepr., Div. Org. Coat. Plast. Chem., 41 (1979) 371.
8. J. K. Gillham, Am. Inst. Chem. Eng. J., 20 (1974) 1066.
9. N. S. Schneider et al., Polym. Eng. Sci., 19 (1979) 304.
10. J. K. Gillham, Polym. Eng. Sci., 19 (1979) 319.
11. J. K. Gillham, ACS Symp. Series, Resins for Aerospace, 132 (1980) Ch. 26.
12. J. K. Gillham et al., in Chemistry and Properties of Crosslinked Polymers, Acad. Press, NY (1977), p. 491.
13. S. J. Washburn, Senior Thesis, Dept. Chem. Eng., Princeton Univ., May 1980.
14. M. J. Doyle, J. K. Gillham, S. T. Washburn and C. A. McPherson, in press.

NETWORK DEVELOPMENT IN FILMS CAST FROM

EMULSIONS OF POLYDIMETHYLSILOXANE

Daniel Graiver, John C. Saam, and Madhu Baile

Dow Corning Corporation
Midland, Michigan 48640

INTRODUCTION

A current trend in the coatings industry is to shift from solvent based systems to emulsions. In elastomeric coatings cross-
linking must occur at some stage during the drying of the latex to
obtain useful films. It is also particularly desirable for cross-
linking to occur at room temperature. This paper describes such a
system based on polydimethylsiloxane (PDMS). It consists of an
anionically stabilized emulsion of high molecular weight PDMS ter-
minated with silanol (1), colloidal silica and dialkytin-dicarboxyl-
ate curing agent (2). Network formation begins at room temperature
before the films are dried and rubbery coherent films form before
all the water evaporates from the system. This paper describes
the cure process and offers an interpretation of the observed
phenomena.

EXPERIMENTAL

Materials

Colloidal silica (Nalco 1115) 15 wt. % solid, particle size 4nm
and sodium silicate solution (44 wt. %) were used without further
purification. Details of the preparation of the emulsion polymer are
to be described elsewhere.

Latex Formulations

Latex formulations were prepared by mixing specified amounts of
the emulsion polymer, the silica or sodium silicate and tin "cata-

227

lyst" adjusting the pH to 10.8 and allowing the system to age 48 hrs.
prior to film casting. Films were cast on PE dishes and aged 14 days
prior to measuring film properties.

Particle Size Measurements

A) Electron Microscopy (EM): Samples were diluted to 0.1 wt. %
solids, dried on a carbon coated grid at room temperature and observed by transmission in a Hitachi HU7 EM. B) Hydrodynamic Chrom-
atography (HDC): Samples were diluted with sodium dichromate water
solution to 0.1 wt. % and the particle diameter was determined by
comparison to a series of monodispersed polystyrene latexes obtained
from Dow Chemical Company.

Stability Measurements

Two series of test tubes were prepared; one series contained
the stable emulsions (.25 gr/15 cc), and the other $La(NO_3)_3$ (10 cc
of various normalties). Mixing was accomplished by pouring the content of two test tubes, one from each series back and forth and the
turbidity of the sol was measured by Hach Laboratory Turbidimeter
Model 2100A. After 24 hours the turbidity of the same sol was measured again and the ratio between the two turbidities was the relative stability.

Measurement of Cross-link Density in Emulsion Particles

The emulsion particles were transferred to heptane by slowly adding 2ml. of the emulsion to a mixture of 20g. anhydrous sodium sulfate in 50ml. heptane. After filtering the sodium sulfate and determining the concentration of polymer the intrinsic viscosity was
determined with an Oswald Fenske viscometer at 25°C.

RESULTS

The cast films when dry do not redisperse in either water or
good solvents for PDMS such as heptane. They swell in both solvents, depending on the amount of silica present (Table I). Typical stress strain curves show a relatively high initial modulus followed by smaller increases in stress at higher strains but when swollen in water the initial modulus decreases (Figure 1). Surface tension of cast films show a high polar component compared to typical
PDMS films filled with silica. Electron microscopy replicas of the
resulting films reveal well defined particle contours even after
three days of drying the films at room temperature. This suggests
a very slow autohesion rate. Excess silica in the dried films can
be observed to form a continuous phase around the original polymer
particles. The particle contours persist for at least 30 days after
film casting and reappear if the film is again wet. Transmission EM

TABLE I. Percent Gel and Swell in Heptane and Water of Films Cast
from Emulsion with 10 parts of colloidal silica after 2 days in
emulsion and 14 days drying at room temperature.

| | Heptane | | | | Water | |
| | As Prepared | | Water Phase Exchanged | | As Prepared | |
ppH Silica	% Gel	% Swell	% Gel	% Swell	% Gel	% Swell
1	49	6800	3	7280	–	–
2	54	5230	7	6010	–	–
5	66	2590	30	5350	>95	70
8	67	1780	39	4330	>95	25
10	68	1370	36	4000	>95	12

of thin sections of films show the original PDMS particles whose
average size is 0.3 μm but whose size distribution is between 0.05
μm and 0.54 μm. The polymer particles are bounded by the much
smaller particles of silica which seems to form the continuous phase

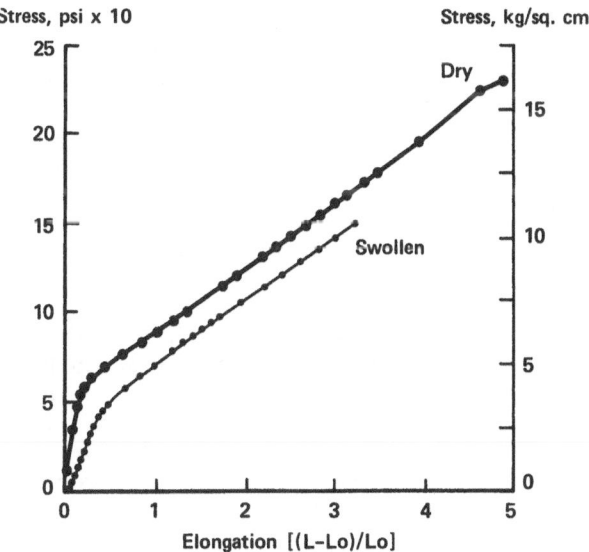

Figure 1:   Stress strain curves of films cast from PDMS emulsion
with colloidal silica (dry), and after swelling in
water (swollen).

This type of morphology, where the hard phase is continuous is not typical of filled elastomers.

One role played by the silica particles appears to be to adhere emulsion particles together in the film (Interparticle Cross-linking). This was demonstrated by exchanging the water phase of the formulations with a Serum Replacement cell (3) to remove un-bound smaller particles of colloidal silica. Films still formed from the retained PDMS emulsion in the cell. These still showed a gel fraction in heptane but it was significantly less than the same films where particles of silica were present (Table I).

The emulsified particles of polymer in these systems had cross-linked to a significant degree within the silicone particles prior to casting films (intra-particle cross-linking). This was established qualitatively by noting the relative distortion of the soft emulsion particles in HDC. The apparent size of the same PDMS emulsion particles increased proportionately as they were cross-linked with increasing doses of γ-radiation even though their corresponding electron micrographs were identical (Figure 2). A similar phenomena was seen in a latex formulation where the colloidal silica was replaced with sodium silicate solution. Here the apparent particle diameter increased with increasing silicate eoncentration. These gave similar but much weaker films than those made with silica. These phenomena can be explained by the distortion of the soft particles under the shear flow in the column rendering the particles a smaller cross-sectional area thus smaller diameter. An increase in the cross-link density of these particles, increases their rigidity which then reduces the distortion **retaining** their shape closer to the original sphere, and the apparent diameter closer to the true diameter.

A better method for calculating the intraparticle cross-link density was based on the relative changes in the hydrodynamic volume of a cross-linked particle (microgel) suspended in an organic solvent. By transferring the polymer particles from emulsion to heptane, the restriction seen in aqueous emulsion on swelling by the surface tension at the oil water interface (4,5) is eliminated. A non cross-linked particle swells until it dissolves, while a cross-linked particle swells until its volume is limited by its network density. The change in the hydrodynamic volume of the microgel determines the intrinsic viscosity of the organosol (6,7) (Figure 3). This change of volume was calculated from Maron's theory (8) for solvent-permeable polymer spheres in solution. Intraparticle cross-links were determined from the extent of swelling of the microgels (9).

The emulsion polymer showed enhanced stability towards coagulation with ionic species after it was formulated (Figure 4) that an additional type of stabilization, perhaps steric, was introduced after the system was formulated.

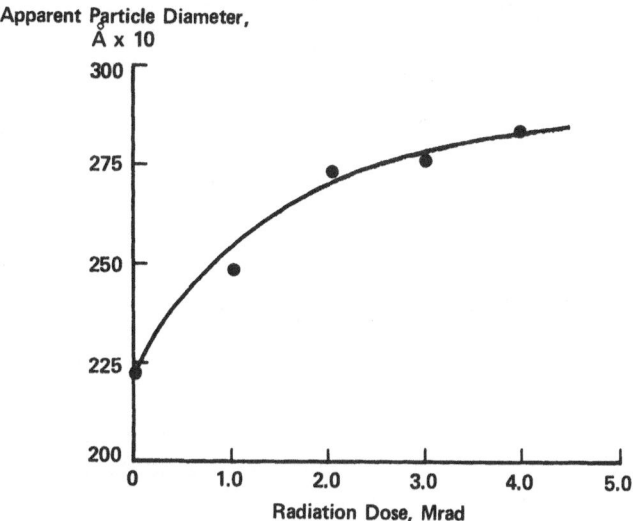

Figure 2:   Variation of the apparent particle diameter in HDC with
increasing radiation dosage in emulsions cure by γ-rays.

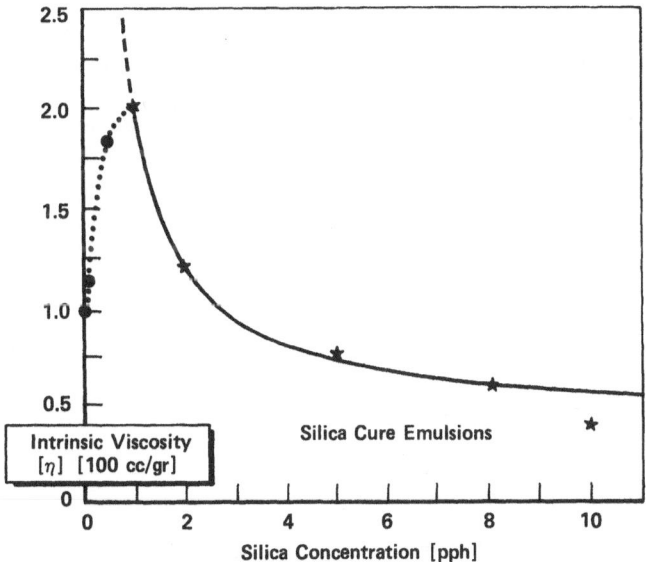

Figure 3:   Intrinsic viscosity of organosols (heptane) with
different levels of cross-linking induced by the
interaction of colloidal silica with PDMS particles
in emulsion.

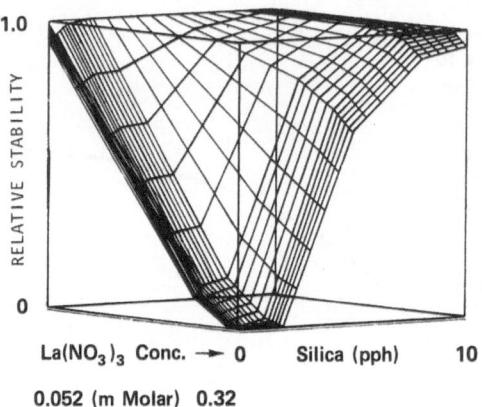

Figure 4: Relative stability of PDMS emulsion containing various
           levels of colloidal silica and La(NO$_3$)$_3$ at neutral pH.

CONCLUSIONS

    The results are consistent with a mechanism where the amphiphil-
lic organotin compound adsorbs at the oil-water interface. The organotin compound can complex with silicates in the water phase either

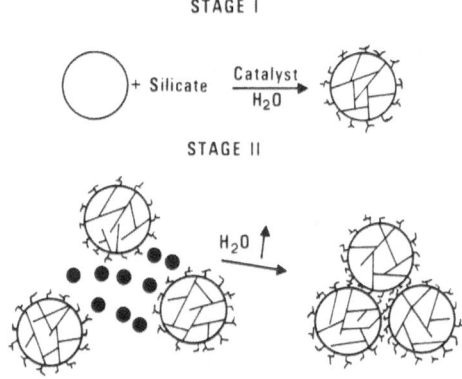

Figure 5: Schematic representation of the cure mechanisms in the
           water phase (Stage 1) and upon film formation (Stage II).

prior to adsorption or at the interface. At the interface the poly-
silicates graft to the silanol at the ends of the polymer. Some
grafts migrate to the interior of the particle to form intra-particle
cross-links while others remain at the surface to provide inter-
particle cross-links to silica or other polymer particles. The latter
takes place as the water evaporates and inter-particle diloxane or
hydrogen bonds rapidly form. The morphology of films cast from such
a suspension would be a continuous hard hydrophilic phase of silicates
surrounding particles of cross-linked polymer (Figure 5). This will
explain (1) a high initial modulus in cast films, (2) films which
swell but do not dissolve both in water and hydrophobic solvents,
(3) a cure as soon as water is removed, (4) cured emulsion particles
prior to film casting, and (5) apparent steric stabilization of the
emulsion.

REFERENCES

(1) J. C. Saam and D. J. Huebner, in preparation.

(2) J. C. Saam, R. D. Johnson and C. M. Schmidt, U. S. Patent
     4,221,688 (1980).

(3) S. M. Ahmed, M. S. El-Aasser, G. H. Pauli, G. W. Pohelein and
     J. W. Vanderhoff, J. Coll. and Interface Sci., 73(2), 388 (1980).

(4) M. Morton, J. Coll. Sci., 9, 300 (1954).

(5) J. Ugelstad, Macro. Chem., 179, 815 (1978).

(6) L. H. Cragg and J. A. Manson, J. Pol. Sci., 8, 265 (1952).

(7) T. C. Chou and A. Rudin, J. Pol. Sci., 11, 2591 (1973).

(8) S. H. Maron, N. Nakajima and I. M. Krieger, J. Pol. Sci.,
     37, 1 (1959).

(9) P. J. Flory, "Principles of Polymer Chemistry", Cornell U.
     Press, Ithaca, New York, 1963, p. 580.

THE EFFECT OF MICROCRAZING ON FATIGUE CRACK

PROPAGATION IN POLYMERS

A. Chudnovsky, A. Moet, I. Palley, and E. Baer

Department of Macromolecular Science,
Case Institute of Technology, Case Western
Reserve University, Cleveland, Ohio 44106

ABSTRACT

A generalized theory of fatigue crack propagation in polymers
is outlined. The theory accounts for fatigue crack propagation
through root craze extension accompanied by simultaneous dissemina-
tion of microcrazing around the crack-root craze system thereby
describing a crack-craze zone (CCZ). In addition to the conventional
crack length, the width of CCZ is introduced as a new internal
parameter. Applying a special version of the second law of thermodynamics: the principle of minimum thermodynamic forces, these
internal parameters are formally described in terms of the reciprocal thermodynamic forces. The rate of crack extension per cycle
was found to depend strongly on changes in the width of CCZ. Results of the model are applied to fatigue crack propagation data
in polystyrene under various loading conditions and a good description of growth rates is observed.

INTRODUCTION

There has been a remarkable effort to elucidate the behavior
of polymers under fatigue loading conditions. Recently, the phenomenology of polymer fatigue has been thoroughly reviewed by
Hertzberg and Manson [1]. In analogy with the more well-studied
low molecular weight materials, fatigue failure in polymers occurs
due to the accumulation of damage resulting from repeated local
plastic deformation. It is generally understood that the formation
of crazes and associated cracks is responsible for the ultimate
failure of many glassy polymers. However, the exact nature of the
mechanism involved is not yet well understood.

In view of the obvious importance that affiliates with the
prediction of Fatigue Crack Growth (FCG) rates, various empirical
as well as analytical approaches have been adopted in an attempt to
develop the law of FCG. Conventionally, FCG "laws" are established
by expressing the propagation rate of a starter crack in terms of
the applied stress or a related function. It is not, therefore, surprising to find several such empirical "laws" in the form: $\dot{\ell} = f(K)^m$,
where ℓ is the rate of crack extension, K is the stress intensity
factor, and m is a numerical factor.

As demonstrated by numerous examples in the recent literature,
the conventional approach suffers from the lack of applicability
to experimental data. This is due to the fact that these developments rest upon the classical view of fracture mechanics. That is,
the failure process involves only the propagation of a single crack-
cut; with or without a plastic zone ahead of its tip. However direct observations in polymers [2,3] and other materials [4] show
that the crack propagates through a damage zone. Specifically,
Bevis and Hull [3] have shown that the application of a tensile
stress to a precracked specimen of polystyrene produced a halo of microcrazes surrounding the slowly growing root crack (Fig. 1a). A
similar zone was also observed under low cycle fatigue of thin samples of polycarbonate and polystyrene [5]. The formation of a zone
of microcrazes around the crack tip could be a source of substantial
energy dissipation. This can be considered analogous to the classical crack-tip plastic zone. Clearly, a major alteration in the
law of crack propagation would be expected due to this phenomenon.

In this paper, we present a more precise analysis of the problem of fatigue crack propagation. The treatment takes into account
the role played by the microcrazing zone and its effect on the
fatigue crack behavior under sinusoidal loading conditions. A more
general expression of the failure process under such loading conditions is formulated. The model provides a good description of
experimental data on the crack growth rate in polystyrene for which
other models proved inadequate.

THEORETICAL DEVELOPMENTS

The Crack-Crazing Zone (CCZ): Based on the observations outlined in the introduction, failure processes in polymers, particularly under fatigue loading conditions, involved crack propagation,
root craze extension, and the simultaneous spread of microcrazes
around the crack-root craze (Fig. 1a). The "crack" system, thus
described, is coined as "Crack-Craze Zone" and is abbreviated CCZ.

Along an axis perpendicular to the CCZ direction, microcrazing
density assumes a bell-shaped distribution. However, for simplicity,
the microcrazing distribution is approximated by an area of uniform

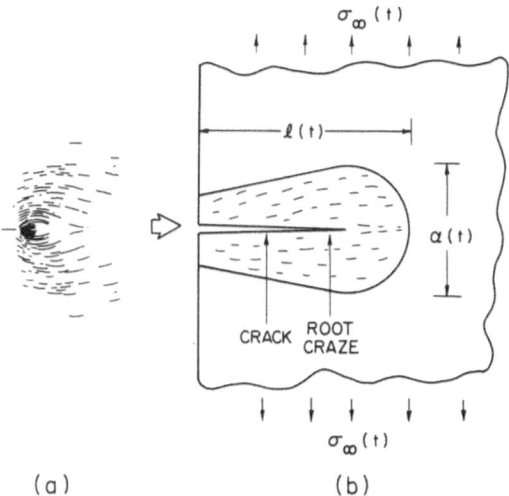

Figure 1.   (a) Composite micrograph showing microcrazing associated
            with crack growth in polystyrene reproduced from ref. [3],
            and (b) Model of the crack-crazing zone (see text).

microcraze distribution with a distinct boundary. The assumed
boundary separates the microcrazed zone from the initial material
(Fig. 1b). We consider CCZ propagation through a thin polymer sample so that two dimensional analysis may be applied and that isothermal conditions may be considered. Two geometrical parameters of
CCZ are accounted for: length (ℓ) and width (α) (Fig. 1b). Description of the CCZ history, therefore, requires the definition of the
functions $\ell = \ell(t)$ and $\alpha = \alpha(t)$, where t stands for time.

Thermodynamics of CCZ Propagation: Viewed in thermodynamic
terms, the propagation of CCZ is an irreversible process which is
best treated by methods of irreversible thermodynamics. The thermodynamic potential G (Gibbs free energy) for the entire system can
be expressed as the sum of two terms

$$G = G_1 \, (\sigma, \, T, \, \ell, \, \alpha) + G_2 \, (\sigma, \, T, \, \ell, \, \alpha) \qquad (1)$$

G_1 and G_2 are the thermodynamic potentials of the CCZ and the complementary part of the body. The stress, σ, and temperature, T,
are external parameters, whereas ℓ and α are internal parameters.

Gibbs free energy, on the other hand, may be resolved into
Helmholtz free energy of the initial material $F_o(T)$, Helmholtz free
energy of deformation F_ε (σ, T) and the work A done by external
forces. Therefore, we may write

$$G_{(k)} = F_{o(k)} \, (T) + F_{\varepsilon(k)} \, (\sigma, \, T) = A_{(k)} \qquad (2)$$

$k = 1$, 2 indicates the CCZ and the complementary part of the body, respectively, because the total displacement $\bar{u}$ may be considered as the sum of a reversible part (elastic) $\bar{u}_e$ and irreversible (plastic) part $\bar{u}_p$ which are associated with corresponding deformations ε_e and ε_p, the work A may be expressed as the sum of an elastic part and a plastic part, i.e.,

$$A_{(k)} = A_{e(k)} + A_{p(k)} \tag{3}$$

By definition, the potential energy (P) associated with the elastic deformation is given by

$$P_{(k)} = F_{\varepsilon(k)} (\sigma, T) - A_{e(k)} \tag{4}$$

From (1), (2), (3) and (4), we obtain

$$G = F_{o1} (T) + F_{o2} (T) + P_1 (\sigma, T, \ell, \alpha) +$$
$$P_2 (\alpha, T, \ell, \alpha) - A_{p1} - A_{p2} \tag{5}$$

The entropy production rate due to propagation of the CCZ, as outlined above, is simply expressed as

$$T \dot{S} = - \frac{\partial G}{\partial \ell} \cdot \dot{\ell} - \frac{\partial G}{\partial \alpha} \cdot \dot{\alpha} \tag{6}$$

Here, $\dot{\ell}$ and $\dot{\alpha}$ are the thermodynamic fluxes conjugated with the thermodynamic forces: $X_\ell = -\partial G/\partial \ell$ and $X_\alpha = -\partial G/\partial \alpha$. These forces may be represented as follows:

$$X_\ell = J + D_\ell - \gamma_{eff} \cdot \alpha \quad \text{and} \quad X_\alpha = M + D_\alpha - \gamma_{eff} \cdot \ell \tag{7}$$

with $J = -\partial P_2/\partial \ell$ being the well known energy release rate, $M = -\partial P_2/\partial \alpha$ is a similar parameter which also has path-independent integral representation for linear media. D_ℓ and D_α are the total dissipation energy rates associated with ℓ and α, respectively. An effective surface energy γ_{eff} consists of the sum of the difference between the free energy density of crazed and uncrazed materials $(f_{o2} - f_{o1})$, the crack surface energy density γ_c and the density of the potential energy of crazed material p_1, i.e.;

$$\gamma_{eff} = f_{o2} - f_{o1} + \gamma_c + p_1 \tag{8}$$

Generally γ_c is a negligible quantity in comparison to the free energy difference $(f_{o2} - f_{o1})$. Due to the small stress in the vicinity of the free edges of a crack, p_1 is also a small quantity.

Conventionally, the law of crack propagation is established by defining the relationship of the thermodynamic force X_ℓ, represented by stress intensity factor K, or the energy release rate J, and critical surface energy density γ, to the conjugated flux described as the crack growth rate ℓ. Because only one kinematic parameter is usually considered, i.e., ℓ, the constitutive law has been always approached by invoking the first law of thermodynamics (by satisfying the energy balance). When two or more kinematic parameters are considered, the energy balance alone would not suffice for the deduction of a constitutive law. In our case, where ℓ and α are considered, an additional principle is necessary.

Therefore, a specific version of the second law of thermodynamics, that is, the principle of minimum thermodynamic forces [7], is considered.

In view of the second law of thermodynamics, the entropy production rate has to be non-negative. In the simple case of $= 0$, we can obtain from (6) and (7):

$$T \dot{S}_i = (J + D_\ell - \gamma_{eff} \cdot \alpha) \dot{\ell} \geq 0 \qquad (9)$$

Consequently, the thermodynamic forces associated with crack growth rate ($\dot{\ell} > 0$) must also be non-negative, i.e.,

$$J + D_\ell - \gamma_{eff} \cdot \alpha \geq 0 \qquad (10)$$

Experimental observations of fracture surfaces [1] clearly show that crack propagates through the root craze by distinct jumps in a discontinuous fashion (discrete crack advance). Between jumps, damage within the craze fibrils accumulates until the system reaches a state of instability at which another jump occurs. Examination of the parallel discontinuous growth bands of many polymers indicates that a single jump approximates the length of a root craze. In this connection, it is believed that a new root craze evolves instantaneously past each jump.

Since both the energy release rate, J, and the effective surface energy γ_{eff} depend only on CCZ configuration, therefore, they will not change between consecutive crack jumps where the system is considered stable. The dissipative energy, on the other hand, accumulates during load excursions. We may therefore, write:

$$D_\ell(\Delta N) = \Psi(\ell, \alpha, \sigma, T) \cdot \Delta N \qquad (11)$$

where Ψ is the energy dissipated per cycle and ΔN is the number of cycles per crack jump.

According to the second law of thermodynamics, crack may grow only when the condition (10) is satisfied, i.e.,

$$J + \Psi \cdot \Delta N - \gamma_{eff} \cdot \alpha \geq 0 \qquad (12)$$

This expression describes necessary but insufficient conditions of instability for the general case. Nevertheless, under the particular conditions considered, where a uniform stress is applied on the boundary of the system, (12) is necessary and sufficient. When instability is reached, i.e., equality of (12) is achieved, CCZ jumps into a new configuration, at which case

$$\Delta N = \frac{\gamma_{eff} \cdot \alpha - J}{\Psi} \qquad (13)$$

As we alluded to earlier, the length of a crack jump ($\Delta \ell$) is considered equal to the length of the root craze. The average speed of CCZ propagation in the direction of the root craze is therefore given by:

$$\frac{\Delta \ell}{\Delta N} = \frac{\Psi \cdot \Delta \ell}{\gamma_{eff} \cdot \alpha - J} \qquad (14)$$

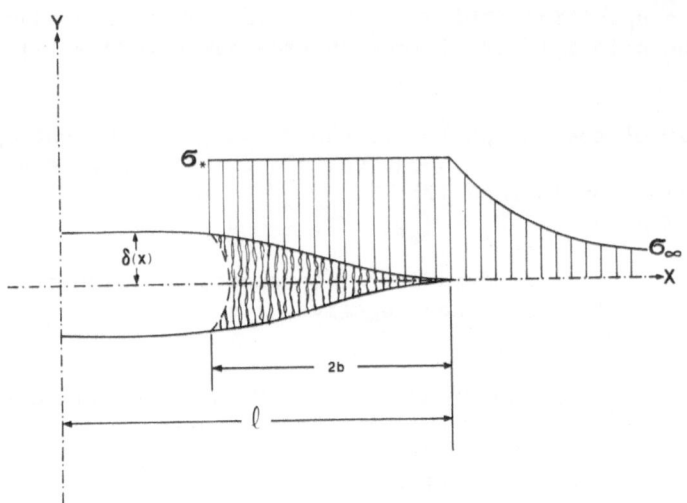

Figure 2.   Crack-root craze configuration according to Dugdale
            model. Note that σ_* replaces σ_y of Dugdale.

In the following section, we proceed to evaluate the independent
variables of this equation.

EQUATION OF CCZ PROPAGATION IN FATIGUE

A.   Energy Dissipation in the Root Craze
     In order to calculate the energy dissipation per cycle in the
root craze Ψ, we introduce few modifications to the Dugdale model
[8] which is illustrated in Fig. 2. In accordance with experimental
observations indicating discontinuous crack growth [6], we suggest
that the root craze does not undergo any significant propagation
in the period between two crack jumps. During this period of no
growth, the fibrils are assumed to be under a finite stress, distributed uniformly and defined as $\sigma_*(t) \neq 0$. The stress distribution assumed (Fig. 2) provides the following relationship between
the crack length (ℓ) and root craze (2b) in terms of external stress
$\sigma_\infty(t)$, and $\sigma_*(t)$:

$$\frac{2b}{\ell} = 2 \sin^2 \left(\frac{\pi \cdot \sigma_\infty(t)}{4 \, \sigma_*(t)} \right) \tag{16}$$

It should be noted that the formalism of the relationship (16) is
identical to that derived by Dugdale [8]. However, the physical
meaning of $\sigma_*(t)$ is different from that associated with σ_y in
Dugdale model as explained later in the discussion.

The microscopic evidence obtained from fatigue crack growth in many polymers justifies the assumption of a constant $2b/\ell$ ratio, within the period between two consecutive jumps. Accordingly, and from (16):

$$\frac{\sigma_\infty(t)}{\sigma_*(t)} = \text{constant} \tag{17}$$

For sinusoidal loading, the time dependent stress may be expressed as the following function:

$$\sigma_\infty(t) = \frac{\sigma_{max}}{2} [(1 + R) + (1 - R) \sin \omega t] \tag{18}$$

where $R = \sigma_{min}/\sigma_{max}$ is the load ratio and ω is the load frequency. From (17) to (18), we may write

$$\sigma_*(t) = \frac{\sigma_{*max}}{2} [(1 + R) + (1 - R) \sin \omega t] \tag{19}$$

Here, σ_{*max} relates to the critical craze initiation stress which formally corresponds to Dugdale's σ_y. This stress is considered as the stress at which a "new" root craze initiates ahead of the crack at the moment of crack jump.

During a load cycle, the craze is believed to maintain the same geometry. Hence, the craze opening displacement is described by the following function:

$$\delta(x,t) = \delta_*(t) \cdot \phi(x) \tag{20}$$

Here, δ_* is the opening displacement at the crack/root craze boundary (Fig. 2). The function $\phi(x)$ expresses the normalized distribution of the craze opening displacement along its axis and is equal to unity at the crack/root craze boundary. To account for the dissipation processes occurring within the craze material, it is reasonable to introduce a phase lag λ of the displacement $\delta_*(t)$ with respect to $\sigma_*(t)$. In accordance with equation (19), we may write:

$$\delta_*(t) = \delta_{*max} \, 1/2 \cdot [(1 + R) + (1 - R) \sin (\omega t - \lambda)] \tag{21}$$

From which the rate of crack/craze opening displacement is

$$\dot{\delta}_*(t) = \delta_{*max} \cdot \frac{1-R}{2} \cdot \omega \cos (\omega t - \lambda) \tag{22}$$

The dissipation energy per cycle Ψ of equation (11) can be given as:

$$\Psi = \int_o^{2\pi/\omega} \int_o^{2b} \sigma_*(t) \cdot \dot{\delta}_*(x,t) \, dx \cdot dx \tag{23}$$

which upon integration yields:

$$\Psi = \beta_1 \, \sigma_{*max} \cdot \delta_{*max} \tag{24}$$

The term β_1 of the above expression is given by:

$$\beta_1 = \frac{(1-R)^2}{2} \sin \lambda \cdot \int_o^{2b} \phi(x) dx \tag{25}$$

In view of recent developments [9] in which the Dugdale crack opening displacement was derived in terms of the stress intensity factor, we may similarly write:

$$\sigma_{*max} \cdot \sigma_{*max} = \frac{K^2_{max}}{E} \tag{26}$$

Finally, the energy dissipation associated with the root craze per cycle can be given by

$$\Psi = \beta_1 \frac{K^2_{max}}{E} \tag{27}$$

B. Dependency of Crack Jump on the Stress Intensity Factor

We may recall that the crack propagates through the root craze by distinct jumps, each such jump is $\Delta\ell = 2b$ (the root craze length). From (16), (17), it is obvious that $\Delta\ell$ is proportional to the crack length ℓ (Fig. 2). In fracture mechanics terms, the stress intensity factor is related to the crack length through the relationship: $K^2 = \sigma^2 \pi\ell V(\ell/B)$, where the function V reflects the dependence of the stress intensity factor on the geometry of the specimen. B is a characteristic scale and is usually taken as the width of the specimen. Consequently the crack length is given by

$$\ell = \beta_2 K^2_{max} \tag{28}$$

where $\beta_2^{-1} = \sigma^2 \pi \cdot V(\ell/B)$. As $\ell/B \leq 1/2$, the function $V(\ell/B) \sim 1$ and the crack length is proportional to K^2.

Using equations (14), (25) and (26) and substituting $J = \frac{K^2_{max}}{E}$, the equation of CCZ propagation may ultimately take the form

$$\frac{\Delta\ell}{\Delta N} = \frac{\beta K^4_{max}}{E \cdot \gamma_{eff} \cdot \alpha - K^2_{max}} \tag{29}$$

where $\beta = \beta_1\beta_2$.

DISCUSSION

The term $E \cdot \gamma_{eff} \cdot \alpha$ of equation (29) can be considered equivalent to the conventional critical stress intensity factor K^2_c. Under conditions where α may vary during crack growth history K^2_c may become a history-dependent parameter.

Three distinct crack growth regimes can be recognized according to equation (29). The first regime is defined by $\frac{K^2}{K^2_c} \ll 1$ at which case (29) can be rewritten as:

$$\frac{d\ell}{dN} = \frac{\beta K^4_{max}}{K^2_c} [1 + (\frac{K_{max}}{K_c})^2 + ...] \sim \frac{\beta K^4_{max}}{K^2_c} \tag{30}$$

Evidently, this regime is well approximated by Paris equation (11). The condition $\frac{K}{K_c} \sim 1$ defines the third regime in which $\frac{d\ell}{dN}$ vs. K

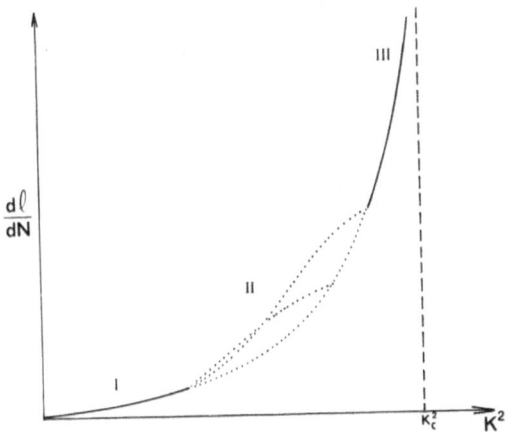

Figure 3. An illustration of the three-regime crack propagation
history according to equation (29).

exhibits an asymptomatic behavior (Fig. 3). Regime II can be quite
complex as shown by the dotted lines in the figure. The complexity
of this regime is strongly dependent on changes of which may be
envisioned negligible in regimes I and III as will be explained
below.

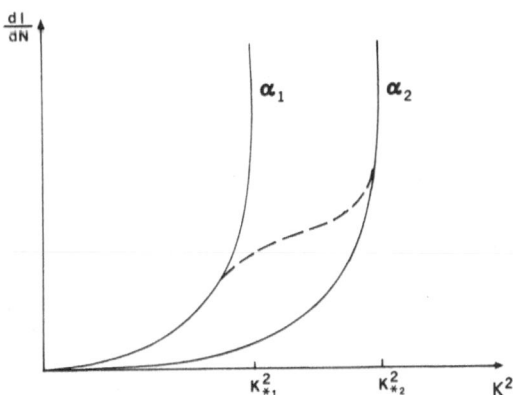

Figure 4. Crack propagation behavior at two different values of α.
The dashes illustrate the transitional behavior when α
changes into a larger configuration.

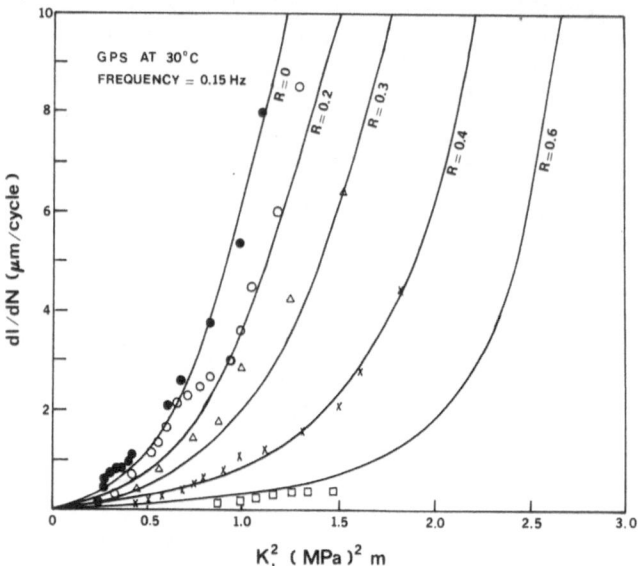

Figure 5. Fatigue crack propagation data for polystyrene taken
          from reference [12] and plotted in terms of the proposed
          model.

Solution of equation (29) for two different values of $\alpha(\alpha_1 < \alpha_2)$
is represented in Fig. 4, where the strong dependence of the crack
growth behavior on the magnitude of α is shown. It is therefore ar-
gued that the critical stress intensity factor, conventionally
thought to be a material property, could be strongly dependent on
the CCZ width and related changes. Such dependence has been established theoretically and experimentally for other materials [10].
In that report, K_c was found to be strongly dependent on the history of loading. It should be emphasized here that the growth of α
represents only one source for changing K_c. The evolution of
another microcrazing zone far ahead from the CCZ constitutes another
influential source to affect the magnitude of K_c. Naturally, the
model proposed here can account for such diverse phenomena. During
the crack propagation history, α changes into a higher value thereby causing reduction in the crack propagation rate or even crack
deceleration. This case is represented by the dotted line bridging
the two constant curves (Fig. 4) such plateau has been repeatedly
observed in several materials including polymers under fatigue
loading conditions. An example is shown in Fig. 5. The observed
deceleration in crack growth strongly indicates the existence of a
mechanism of energy absorption different from the main crack growth.
Elucidation of such mechanism(s) is important for understanding
factors governing material toughness.

APPLICATION TO DATA

As an illustrative test for the model, some fatigue crack growth data [12] are shown in Fig. 5 in the following form of equation (29):

$$\frac{d\ell}{dN} = \frac{\beta K_{max}^4}{K_c^2 - K_{max}^2} \tag{31}$$

Since the width of CCZ was not reported, the term K_c^2 has been approximated as an average of $3.0(MPa)^2 \cdot m$ from the recently accumulated data of polystyrene [13]. The dependence of β on R was estimated from the experimental data of figure 5. These values are tabulated below:

R	0	0.2	0.3	0.4	0.6
β	12	7.5	4.0	1.56	0.46

In spite of the crude estimation made to evaluate equation (29), due to the absence of precise measurements of the related parameters, the model provides a reasonable description of the fatigue crack propagation. In a future publication, we report on more accurate measurements of the CCZ zone in various polymers together with further refinement of the model.

ACKNOWLEDGMENT

The authors gratefully acknowledge the generous support of the Office of Naval Research, Contract No. N00014-75-C-0795, and the National Science Foundation, Grant No. DMR 77-24952.

REFERENCES

1. R.W. Herzberg and J.A. Manson, "Fatigue of Engineering Plastics", Academic Press, New York (1980).
2. D. Post, Proc. Soc. Exp. Stress Anal., 12, 99 (1954).
3. M. Bevis and D. Hull, J. Mater. Sci., 5, 983 (1970).
4. R.G. Hoagland, G.H. Han and A.R. Rosenfield, Rock Mechanics, 5, 77 (1974).
5. J. Botsis, A. Moet, A. Chudnovsky, unpublished.
6. Chapter 4 of ref. 1.
7. P. Glansdorff and I. Prigogine, "Thermodynamic Theory of Structure Stability and Fluctuations", Wiley-Interscience, New York (1971).
8. D.S. Dugdale, J. Mech. Phys. Solids, 8, 100 (1980).
9. J.R. Rice, H. Liebowitz, in (Ed.), "Fracture", Vol. II, Academic Press, New York (1968).
10. V. Khanodgen and A. Chudnovsky, "Thermodynamic Analysis of Quasistatic Crack Growth in Creep", in L. Kurshin (Ed.)

"Dynamics and Strength in Aircraft Construction", (Russian) Novosibrisk (1978).

11. P.C. Paris, Paper No. 62-Met-3, ASME, New York (1962).

12. Y.W. Mai and J.G. Williams, J. Mater. Sci., 14, 1933 (1979).

13. G.P. Marshall, L.E. Culver and J.G. Williams, Int. J. Fract., 9, 295 (1973).

ADVANCES IN STRENGTH PREDICTION FOR

SHORT-FIBER REINFORCED PLASTICS

J. L. Kardos

Materials Research Laboratory and
Department of Chemical Engineering
Washington University
St. Louis, MO. 63130

INTRODUCTION

Prediction of strength for short fiber reinforced plastic systems is a complex but industrially crucial problem. Even in the case of unidirectionally aligned fibers with tensile stress applied in the fiber direction, failure may occur in the fibers, in the matrix phase, or at the interface. Furthermore, failure may take place in a tensile or shear mode and may be brittle or ductile in nature.

Historically, the strength of materials approach to composite strength prediction began with the work of Cox (1) in 1952, which, although not exact, served as the basis for later developments of what is now called the "shear lag" analysis. Later detailed analyses by Dow (2) and Rosen (3) produced the same basic result. In 1964 Cottrell (4) formulated the basic shear lag analysis based on a single fiber and introduced the concept of a critical fiber length, ℓ_c, above which the fiber's ultimate strength will be fully utilized. One year later Kelly (5-7) extended this principle to describe the behavior of short fiber reinforced metals and predicted that up to 95% of continuous fiber composite properties should be attainable with short fiber systems if perfect uniaxial alignment is achieved. Unfortunately, perfectly uniaxial reinforcement is usually not attained experimentally. Lees attempted to modify the shear lag approach (8) by dividing the fiber length distribution into those fibers below and at or above the critical aspect ratio, and then summing the contributions to the two different failure modes. This approach also fails to predict experimental reality. Other approaches have been made from a strength-of-materials viewpoint (9), but again these single-fiber analyses fail to describe systems of commercially important reinforcement levels.

The problem is schematically summarized in Figure 1. Note that in passing from continuous to uniaxially aligned short-fibers, the tensile modulus is maintained whereas the strength drops precipitously. For a randomly oriented sample, the modulus drops as expected and the strength drops once more to a value about 1/4 of the continuous unidirectional value for the same fiber volume loading. The key to the initial drop in strength for uniaxial orientation is the presence of and interaction between fiber ends. It was not until the pioneering work of Chen in 1971 (10) that the interaction between neighboring short fibers was fully taken into account. Chen utilized a finite element approach, which included a distortional energy criterion, to calculate the strength of several uniaxially aligned short fiber systems. He found that the composite strength reached a plateau as the fiber aspect ratio was increased at constant volume loading. Furthermore, this plateau occurred at only 80% of the continuous fiber value for tungsten-copper composites and at an incredibly low 55% and 60% of continuous values for boron-epoxy and glass-epoxy respectively. Barker and MacLaughlin (11) and Riley (12) also concluded that interacting fiber stress concentrations should reach a plateau at large aspect ratios.

In this paper we utilize a new approach which accounts for the large stress concentration penalties in a perfectly aligned short fiber composite (13). Although empirical, the method permits calculation of a strength reduction factor which can then be utilized with an appropriate failure criterion to calculate the strength of a wide range of short fiber composite systems.

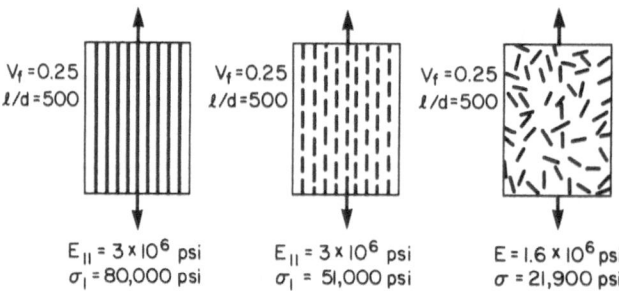

Figure 1. The effect of fiber length and orientation on the mechanical properties of glass fiber epoxy composites.

RESULTS

Calculational Format

To predict the strength of short fiber systems, it is advantageous to use a laminate analogy. In this approach, the short-fiber reinforced part is thought of as being composed of several plies or layers, each containing uniaxially aligned short fibers. The plies are oriented in the laminate to replicate the actual part fiber orientation distribution and the linear stress-strain properties calculated by analyzing the individual ply responses to the overall applied stress. The general procedure will now be outlined.

1. The first step is to calculate the plane stress moduli of the individual uniaxially oriented plies and then transform them to the particular orientations occupied by the plies in the laminate. These transformed moduli are then summed through the layers of the laminate thickness and the overall effective engineering moduli of the laminate calculated. If the part being analyzed is a random-in-a-plane sheet, there will be eight plies, two at each of 0°, $+45^\circ$, -45°, and 90° angles, respectively. For a non-random orientation, the number of plies and their thicknesses will depend on the angular increments chosen to divide up the distribution. The necessary equations and an example calculation for random-in-a-plane orientation are described in detail in reference 14.

2. The second step requires the choice of a failure criterion for a continuous fiber-reinforced ply having uniaxial orientation. One such criterion holds that if the ply is strained beyond certain maximum values in tension and shear, the ply will fail. This maximum strain criterion has been shown to work quite well and will be used in this format; however, any other failure criterion could alternatively be chosen. Experimental values for the allowable maximum strains must be determined for the system of interest for transverse (to the fibers) tension, longitudinal tension, and in-plane shear.

3. Next, the allowable strains for the continuous fiber ply must be reduced to account for the short fiber penalty. This is conveniently accomplished using the Halpin-Kardos Equations which yield a strength reduction factor (SRF) which, along with the moduli of the discontinuous and continuous systems, permits calculation of the allowable strains for the short fiber system.

4. For each ply at its particular orientation in the laminate, the applied laminate strains are transformed from the axial direction of the laminate to the principal directions of the ply. The smallest axial strain that will cause failure of the ply is noted and the order of the ply failure is thus determined. Once the first-ply failure axial strain is known, this is assumed to be the

midplane strain of the laminate and the stress causing that failure
can be calculated.

5. The calculation continues by deleting the failed ply from
the laminate and recalculating the overall laminate stiffness. The
next lowest transformed axial strain allowable is chosen and the
incremental stress needed to produce this strain is calculated. The
second failed ply is deleted and the calculation is repeated. After
all the plies have failed, the incremental strains and stresses are
summed to obtain the ultimate stress and strain of the laminate.

The Short Fiber Penalty

In examining Chen's work and our own laboratory results on uniaxially aligned short fiber systems, two important characteristics
are clear. One is that a plateau in strength is reached at large
fiber aspect ratios when the fiber volume fraction is held constant.
The aspect ratio, at which the knee in the curve appears, depends on
the type of fiber and matrix and is much larger than that predicted
by the shear lag analysis. Secondly, if one plots strength versus
aspect ratio at constant volume fraction of fibers for different
systems, a family of sigmoidal curves results, whose plateau values
for strength are much lower than those predicted by shear lag analysis and whose values also depend on the particular fiber-matrix
combination used. If one could collapse all these curves on a single
master curve utilizing generalized normalization parameters, then
the stress concentration penalty for interacting fiber ends could
be accurately determined for any short fiber system.

The problem may be approached by considering how the curves may
be shifted both horizontally and vertically to produce a normalized
master curve. In effect, use may be made of an aspect of dimensional
analysis in which division of one dimensionless group by another can
collapse the data if the groups are properly chosen.

To accomplish the vertical shift, a strength reduction factor
[SRF] is defined as the uniaxially aligned short fiber system
strength divided by the strength of an aligned continuous fiber
system of the same volume fraction fibers. Thus,

$$[SRF] = \bar{\sigma}_c / \sigma_R V_R \tag{1}$$

wherein the matrix contribution to the rule-of-mixtures continuous
fiber strength is neglected. σ_R is the fiber strength and V_R the
volume fraction reinforcement. As the aspect ratio approaches
unity, the [SRF] approaches that for a sphere-filled system, namely
$[SRF]_o$. This lower bond can be evaluated by utilizing the results
of Narkis et al. (15-17) for glass bead-filled thermoplastics and
thermosets,

$$[SRF]_o = \frac{\sigma m \frac{E_c}{\sigma_R V_R E_m}(1-V_R^{1/3})}{\sigma_R V_R E_m} ,$$ (2)

where σ_m is the matrix strength.

The value for the [SRF] at large fiber aspect ratios, $[SRF]_\infty$, may be shifted vertically by utilizing the fiber-to-matrix stiffness ratio, E_R/E_m, which in fact controls the magnitude of the stress concentrations at the fiber ends (10-12). The best fit for this shift yields

$$[SRF]_\infty = 0.5 + (E_R/E_m)^{-0.87} \text{ for } E_R/E_m > 5$$ (3)

The horizontal shift parameter, β, can be developed by noting the parameters of importance in the shear lag analysis. The critical aspect ratio and the ratio of fiber strength to matrix shear strength are related by a constant. Thus the horizontal shift parameter is

$$\beta = (\ell/d)/(\sigma_R/\tau_m)$$ (4)

where τ_m is the shear strength of the interface or matrix, whichever is lower.

The final equation for the master curve may be normalized and expressed as follows:

$$G = \frac{[SRF] - [SRF]_o}{[SRF]_\infty - [SRF]_o} = 1 - 0.97 \exp[-0.42\beta]$$ (5)

The above equations, collectively denoted as the Halpin-Kardos Equations, predict the lowering of continuous fiber strength due to the discontinuous nature of the reinforcement.

Random-in-a-plane Fiber Orientation Distributions

Following the format outlined above, the uniaxial tensile strength of a random-in-a-plane orientation, E-glass/epoxy composite may be predicted. The calculational details have been presented elsewhere (14) and only the salient results are summarized here. The Halpin-Kardos equations have been made somewhat more precise since the calculations in reference 14 were made, but the strength results are the same. For a 50 v % fiber loading the stress-strain results are summarized in Table 1 below. Note that the 90° ply failed first, followed by the $+45^o$ plies, and then the 0° ply. Note also that as the plies fail, the laminate stiffness decreases yielding a piecewise linear stress-strain curve. While the experimental curve does not look like this, the ultimate stress and strain agree well with experiment. The results are shown in Figure 2 for a wide range in volume loading. The open circles represent the experi-

Table 1 – Summarized Stress-Strain Data

Ply Failure	ε_x	$\bar{E}$	$\Delta\varepsilon_x$	$\Delta\sigma_x = \bar{E}\Delta\varepsilon_x$	σ_x
90°	0.0038	2.74×10^6	0.0038	10,280	10,280
$\pm45^\circ$	0.0141	2.32×10^6	0.0104	24,010	34,290
0°	0.0155	$1.4 \ \times 10^6$	0.0014	1,960	36,250

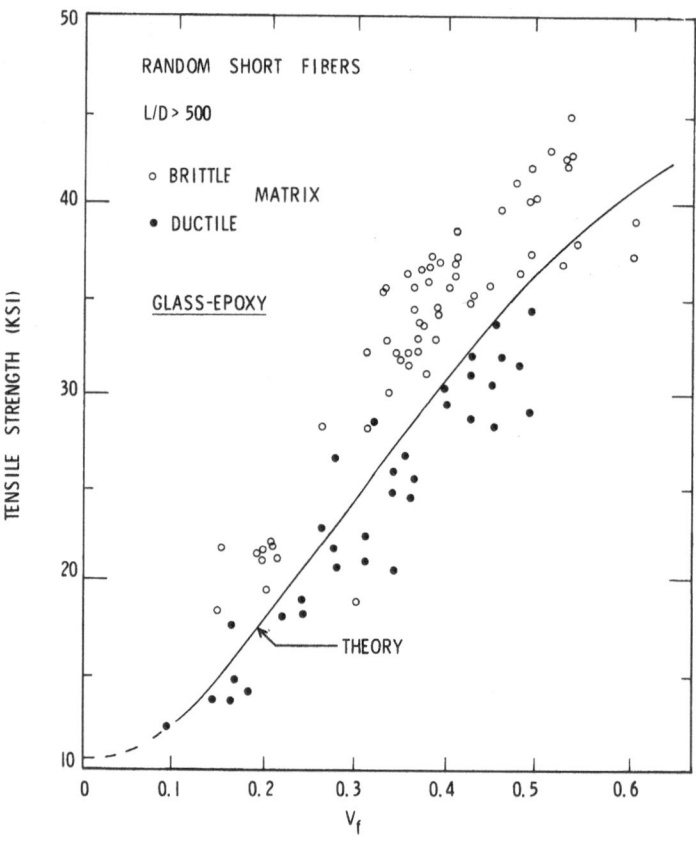

Figure 2.   The effect of volume fraction fibers on the tensile
strength of glass epoxy composites for a random short
fiber system.

mental points for which the predictions were made. Clearly, the prediction is conservative, even when the experimental scatter is considered; but the prediction is well within 20% of reality, which
places it within workable design bounds.

Biased Fiber Orientation Distributions

Many injection-molded parts contain non-random, partially oriented fiber orientation distributions. To predict strength for
these systems, exactly the same format can be used. The laminate is
divided into weighted groups of angle-plies to match the actual fiber
distribution and the order of ply failure is determined as before by
transforming the laminate strain to the principal directions of each
particularly oriented ply.

A carefully characterized data base for an E-glass/ductile epoxy
matrix system has been described elsewhere (18) and was used to as-
sess the accuracy of the strength prediction approach for a biased
orientation. Three different fiber bundle lengths were processed to
yield partially aligned systems whose fiber orientation distributions
were accurately determined. The calculational format for strength
was applied (see reference 18 for details), utilizing the appropriate constituent properties for the ductile epoxy matrix. Table 2
compares the experimental and predicted results for three different
fiber bundle aspect ratios at a fiber volume loading of 50%.

For the lowest aspect ratio, the calculation is again conservative by about 10%, whereas at the higher aspect ratios the predicted
value over-estimates the strength by about 10%. The experimental
values represent the mean of 10 test specimens. Clearly, the agreement here is even better than in the case of random orientation.

Table 2 - Comparison of Predicted with Experimental
Ultimate Tensile Strength and Strain

Fiber Bundle Aspect Ratio	Experimental		Predicted	
	$\bar{\sigma}_x$, ksi	ε_x, %	$\bar{\sigma}_x$, ksi	ε_x, %
185	60.9	1.45	55.6	1.32
398	65.0	1.49	71.0	1.64
557	74.0	1.57	79.0	1.66

CONCLUSIONS

The laminate analogy in conjunction with a maximum strain failure criterion and an empirical strength reduction factor form a
calculational format which can be used successfully to predict the
ultimate strength and strain of short fiber composites. The method
accounts for the fiber orientation distribution, fiber aspect ratio,
fiber volume loading, constituent properties and, most importantly,
the stress concentration penalty caused by the fiber ends. For both
random-in-a-plane and partially aligned fiber orientation distributions, the approach yields agreement with experiment within the 20%
needed for preliminary design accuracy.

ACKNOWLEDGMENT

This work was supported by the Polymers Program of the National
Science Foundation under Grant No. DMR78-12806.

REFERENCES

1. H. L. Cox, Br. J. Appl. Phys. 3:72 (1952).
2. N. F. Dow, General Electric Report R635D61 (1963).
3. B. W. Rosen, Mechanics of composite strengthening, in "Fiber
    Composite Materials", American Soc. Metals, Metals Park,
    Ohio (1965).
4. A. H. Cottrell, Proc. Roy. Soc. Series A, 282A:2 (1964).
5. A. Kelly, Fibre Reinforcement, in "Strong Solids", Clarendon
    Press, Oxford (1969).
6. A. Kelly and W. R. Tyson, J. Mech. Phys. Solids. 13:329 (1965).
7. A. Kelly and G. J. Davies, Metallurgical Rev. 10:1 (1965).
8. J. K. Lees, Polym. Eng. Sci. 8:195 (1968).
9. J. O. Outwater, Jr., Mod. Plast. 33:156 (1956).
10. P. E. Chen, Polym. Eng. Sci. 11:51 (1971).
11. R. M. Barker and T. F. MacLaughlin, J. Comp. Mater. 5:492
    (1971).
12. V. R. Riley, J. Comp. Mater. 2:436 (1968).
13. J. L. Kardos, J. C. Halpin, and S. L. Chang, "Rheology, Vol. 3 -
    Applications", G. Astarita, G. Marrucci and L. Nicolais,
    eds., Plenum Press, New York (1980), pp. 255-260.
14. J. C. Halpin and J. L. Kardos, Polym. Eng. Sci. 18:496 (1978).
15. R. E. Lavengood, L. Nicolais, and M. Narkis, J. Appl. Polym.
    Sci. 17:1173. (1973).
16. M. Narkis, Polym. Eng. Sci. 15:316 (1975).
17. M. Narkis, J. Appl. Polym. Sci. 20:1597 (1976).
18. E. Masoumy, L. Kacir and J. L. Kardos, Polym Composites, in
    press.

THERMALLY STIMULATED CURRENT OF SOLID PULLULAN

K. Nishinari, D. Chatain, and C. Lacabanne

Laboratoire de Physique des Solides
Associé au C.N.R.S.
Université Paul Sabatier
118 Route de Narbonne
31062 Toulous Cédex (France)

INTRODUCTION

Pullulan is used as food material and as food packaging films. Nevertheless, its physical properties have not been investigated in detail. In order to develop further utilization (1), it is desirable to know the relation between its chemical structure and physical properties.

Moreover, since polysaccharides such as cellulose, amylose, dextran and pullulan have the same structural units, they can be used as models for studying the influence of the various modes of linkages on the relaxational behavior of polymeric solids. However, the electrical and mechanical properties of solid polysaccharides have been studied by only a few authors (2-8). Recently, a great interest has been devoted to the binding of water on biopolymers (9-11). Nevertheless, a lot of problems remain unsolved. In order to shed some light upon these points, Thermally Stimulated Currents (TSC) of pullulan films have been investigated. The TSC technique which has been applied with success in polymer physics (5,12-55), seems to be well adapted to this study.

MATERIALS

The chemical structure of pullulan is represented on Figure 1. Pullulan film, 30μ thick, which was kindly supplied by Hayashibara Biochemical Laboratories Inc., was dissolved in pure water and cast into a film of 250μ thickness on a teflon plate. The average mol-

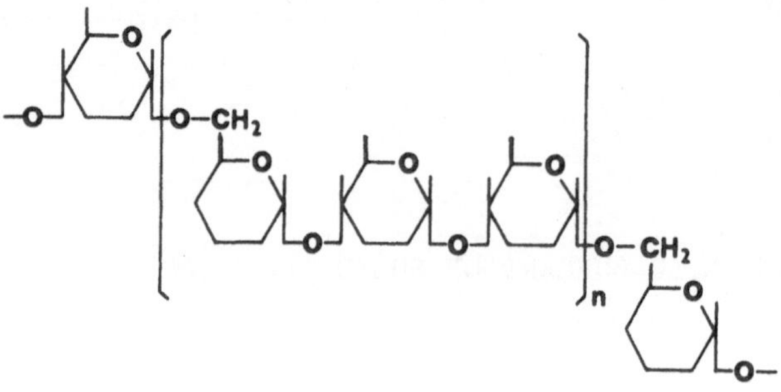

Figure 1. Structure of pullulan.

ecular weight and the density d at 60 R.H. measured in the above
laboratory were 300,000 and 1.20 respectively. Gold electrodes
were deposited in vacuo onto both surfaces of the film. The temperature of the film during evaporation being 30°C, the morphology of
the film was not modified during evaporation.

EXPERIMENTAL

     The TSC method is as follows: the sample is first subjected
to a given electric field E at a temperature T_p for a time Δt: T_p
and Δt are chosen to allow complete orientation of the mobile units
that one wishes to consider. The temperature is then lowered to
$T_o \ll T_p$ where any molecular motion is completely hindered: then the
field is taken off. The sample is subsequently warmed at a controlled rate. The mobile units can reorient. The current density j
and the temperature T are recorded versus time. Then, we obtain
"global" TSC spectra. In pullulan as in all polymers they are
complex: they have been experimentally resolved by using the technique of fractional polarizations (9).

GLOBAL TSC SPECTRUM

     Global TSC spectra of a pullulan film for various moisture
contents are shown in Figure 2. Three peaks were observed at about
50°C, -50°C and -130°C.

Peak at 50°C
     This peak is seen to decrease in magnitude and shifts to higher
temperatures upon drying. This behavior might be explained by the

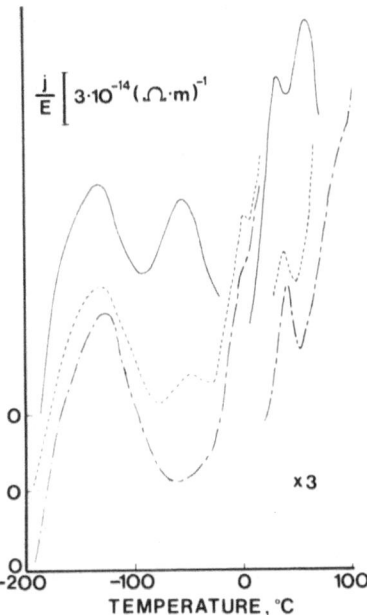

Figure 2.   Global TSC spectra of pullulan film with different mois-
            ture content. ——— heated at 100°C, cooled slowly and
            kept for 5 hours in vacuo (10^{-5}-10^{-6} mm Hg). — — heated
            at 120°C, cooled slowly and kept for 5 hours in vacuo.
            — — heated at 160°C, cooled slowly and kept for 27 hours
            in vacuo.

presence of water molecules acting as plasticizer and facilitating
dipolar movements. This hypothesis will also explain the evolution
of the thermally stimulated creep peak observed for the same material
in the same temperature range (56) and the broad endotherm exhibited
by the hydrated sample over the temperature range 20-50°C (57).

    The shoulder of the depolarization current observed at about 0°C
for hydrated specimens was considered to be due to the fusion of
ice.

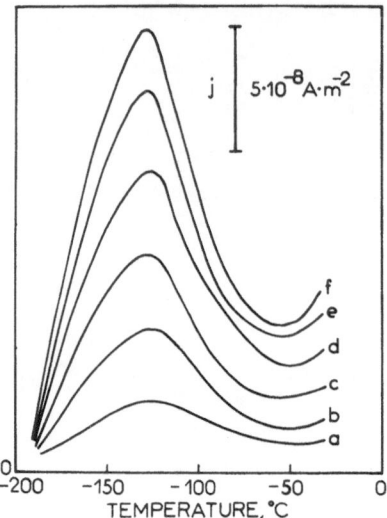

Figure 3. Electric field dependence of the TSC peak at -130°C in pullulan film. a. 0.2×10^{6} V/m; b. 0.4×10^{6} V/m; c. 0.6×10^{6} V/m; d. 0.8×10^{6} V/m; e. 1.0×10^{6} V/m; f. 1.2×10^{6} V/m.

Peak at -50°C
    This peak is not observed in highly dehydrated samples; its magnitude increases when the specimen absorbs a slight moisture. This process has been related with bound water; it has also been observed by dynamic mechanical and dielectric relaxation not only in pullulan (58) but also in amylose (8).

Peak at -130°C
    In hydrated films, this peak is masked by the steep rise of the depolarization current. For slightly hydrated films, the magnitude of this peak is not changed so much by variations of water content.

    In order to ascertain the origin of this process, a series of TSC spectra have been observed in highly dehydrated films for various polarizing field strengths: the variation of the area of the depolarization band versus the polarizing field strength is linear (cf. Figure 3). This suggests that the dipolar orientation is responsible for this process (59).

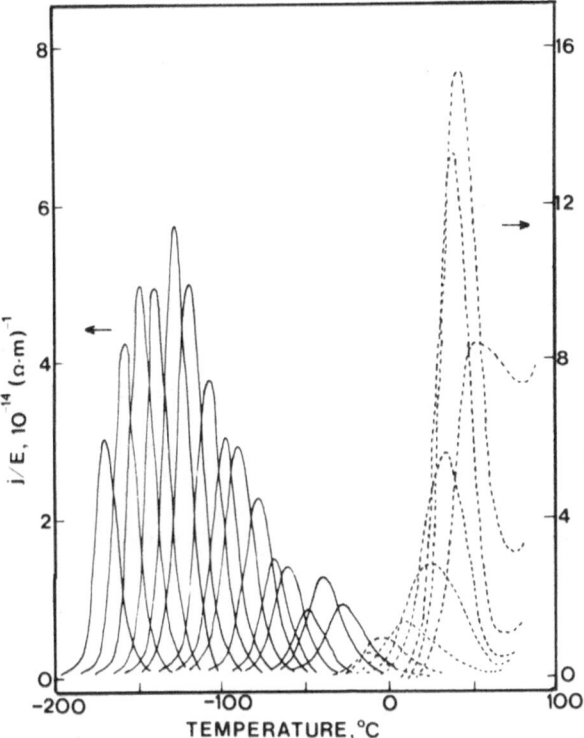

Figure 4.   "Elementary" TSC spectra isolated by the technique of
fractional polarizations in dry pullulan.

In order to verify this hypothesis, two estimations of the
number of relaxing unit N, have have been done:

- from the relationship: $N = N_a d/M$ where N_a is the Avogadro
number, d is the density, M is the molecular weight of the glucosidic residue;

- from the equation giving the polarization of free rotating
dipoles $P = [N\mu^2 E)/3kT_p]$ where k is the Boltzmann constant, T_p the
temperature of polarization, E the electric field and μ the dipole
moment. The value of P is deduced from the area under the TSC peak.

It is important to note that we obtained from both evaluations an average value of $\sim 10^{28}$ dipoles/m^3. This agreement confirms the previous assumption of a dipolar origin for this low temperature relaxation.

ELEMENTARY TSC SPECTRA

The fine structure of complex relaxation modes observed in highly dehydrated films, has been investigated by using the technique of fractional polarization (6): the temperature of polarization was varied in the temperature range - 185°C to +55°C and the temperature window was 10°. The "elementary" processes obtained in that way are shown in Figure 4. The solid line spectra and the left hand scale correspond to the peaks observed between -167°C and -28°C - low temperature mode - while the dashed line spectra and the right hand scale are for the mode between -11°C and +53°C - room temperature (RT) mode. The i-th "elementary" processes can be described in the assumption of a single relaxation time following an Arrhenius equation

$$\tau_i = \tau_{oi} \exp(E_i/kT) \tag{1}$$

where τ_{oi} is the pre-exponential factor and E_i the activation energy.

The semi-logarithmic variation of τ_i versus the reciprocal temperature is plotted on Figure 5 for the dashed line spectra of Figure 4 - RT mode - and on Figure 6 for the solid line spectra of Figure 4 - LT mode. The parameters τ_{oi} and E_i characterizing each elementary peak situated at T_{mi} are reported on Table 1.

The "elementary" processes constituting the RT mode have activation energies varying from 1.1 eV to 0.63 eV (25.3 to 14.5 kcal/mole). This mode whose magnitude is strongly dependent upon the water content of the sample, may be attributed to movements of water molecules introducing during the hydration of the sample.

For the LT sub-mode observed at -50°C, the activation energies of the "elementary" processes take values from 0.51 eV to 0.45 eV (11.7 to 10.3 kcal/mole). The elementary processes constituting the LT sub-mode observed at -130°C, have activation energies varying from 0.35 eV to 0.16 eV (8 to 3.7 kcal/mole); the temperature position of this peak shows that it may correspond to the γ peak observed in the same material at -100°C and at 10 Hz in dielectric and mechanical relaxation (58). In order to perform a better comparison of TSC and relaxation data, the dielectric energy loss has been calculated as a function of temperature and frequency from TSC analysis.

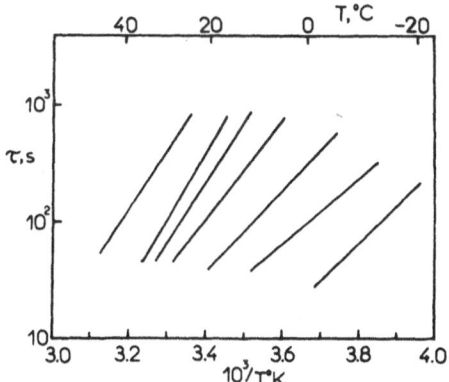

Figure 5. Arrhenius plot corresponding to the dashed line spectra of Fig. 4 -RT mode of pullulan-.

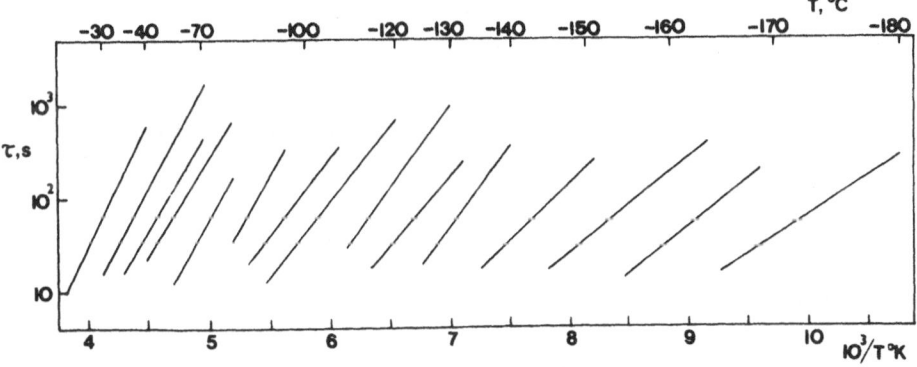

Figure 6. Arrhenius plot corresponding to the solid line spectra of Fig. 4 -LT mode of pullulan-.

Table 1

T_{mi} (°C)	E_i (eV)	$\tau_{o,i}$ (sec.)
− 167	0.16	3.6×10^{-7}
− 156	0.20	4.8×10^{-8}
− 147	0.20	2.4×10^{-7}
− 136	0.24	2.3×10^{-8}
− 125	0.33	9.6×10^{-11}
− 116	0.34	2.0×10^{-11}
− 106	0.35	4.2×10^{-10}
− 96	0.32	1.8×10^{-8}
− 88	0.33	2.5×10^{-8}
− 75	0.32	1.2×10^{-7}
− 66	0.47	9.3×10^{-11}
− 59	0.32	1.6×10^{-6}
− 48	0.46	2.3×10^{-9}
− 38	0.50	7.0×10^{-10}
− 28	0.51	1.4×10^{-9}
− 11	0.63	1.6×10^{-11}
− 5	0.66	1.3×10^{-11}
10	0.37	1.8×10^{-12}
24	0.68	8.3×10^{-11}
34	0.82	8.5×10^{-13}
39	1.03	5.8×10^{-16}
44	1.09	7.5×10^{-17}
53	1.01	8.3×10^{-15}

LT mode (rows − 167 through − 28)

RT mode (rows − 11 through 53)

Activation energies (E_i) and preexponential factors ($\tau_{o,i}$) characterizing the Arrhenius relaxation times τ_o (Equation 1) of the
"elementary" peaks isolated in pullulan at T_{mi}.

DIELECTRIC ENERGY LOSSES

From the parameters τ_{oi} and E_i characterizing the various "elementary" TSC peaks, we have worked out the dielectric energy losses
ε'' as a function of temperature and frequency:

$$\varepsilon''(\omega,T) = \sum_i (\varepsilon_S - \varepsilon_\infty)_i \frac{\omega\tau_i(T)}{1+\omega^2\tau_i(T)^2}$$

where ω is the angular frequency, $(\varepsilon_S - \varepsilon_\infty)$ and $_i$ are respectively
the dispersion and relaxation time of the ith process. We have
represented on Figure 7 the variation of ε'' versus temperature for
frequencies between 10^{-3} and 100 Hertz. The isofrequency 10^{-2} Hz is
analogous to the TSC spectrum. The ε'' peaks observed on Figure 7
correspond to the so-called γ process. The frequency of each ε''
peak has been plotted versus the reciprocal temperature of its maxi-

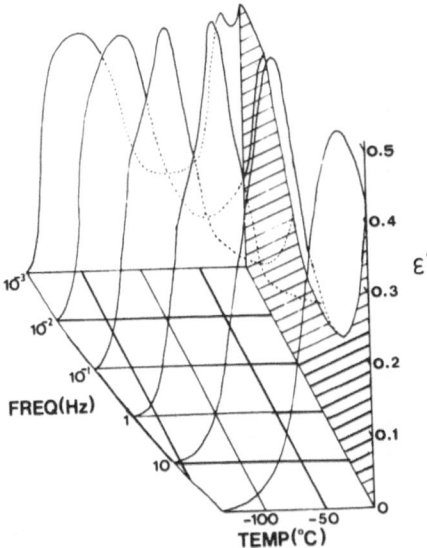

Figure 7.   Dielectric surface deduced from the TSC study of dry
            pullulan.

mum on Figure 8 (filled circles). The "average" activation energy
deduced from the shift of this γ mode is 0.27 eV. For comparison,
it has been reported on the same figure, data obtained by the same
technique on amylose (open circles) (61): as it might be expected
from the fact that the γ TSC peak of pullulan is observed at lower
temperature than the γ TSC peak of amylose, the "average" activation
energy of the γ peak is lower in pullulan than in amylose - 0.33 eV.

     The isofrequencies of 10 Hz have been compared with dielectric
relaxation data performed on the same sample. The γ peak is predicted at a too low temperature. Presently, we have no explanation
of this discrepancy. On the other hand, the TSC peak of pullulan
is found at lower temperature than the TSC peak of amylose: this
datum is in good agreement with the dielectric relaxation one.

DISCUSSION

     Both TSC and dielectric relaxation data show that molecular
movements are easier in pullulan than in amylose. Now, let us

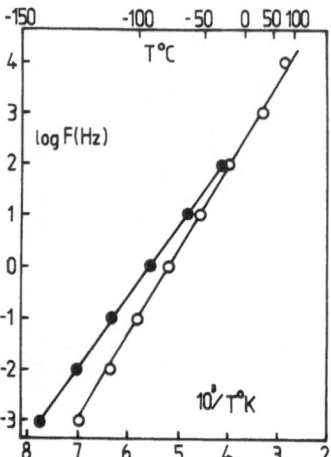

Figure 8.   Arrhenius plot from the dielectric surfaces deduced from
            the TSC study of ● - pullulan (this work) and o - amylose
            (unpublished data).

discuss the possible molecular mechanisms responsible for the γ
mode of pullulan. According to the conformational studies of polymers of D-glucose, the C_1 chair conformation is the most favorable
one in the gluco-pyranose ring (62-63). In the hypothesis of an
isolated pyranose ring, the chair-boat interconversion requires
13.5 kcal/mole (64). In pullulan, the pyranose rings are linked
by α-1,4 and α-1,6 linkages so that the chair-boat interconversion
must require a higher activation energy than for the above case.
Consequently, the chair-boat interconversion cannot be responsible
for the γ process in pullulan.

      Perez et al. (65-67) have calculated from conformational analy-
sis the potential barrier for the rotation of pyranose ring around
the α-1.4 and α-1.6 linkages: they have found 5-6 kcal/mole.
Zhbankov (68) has obtained the same value for the internal rotation
of the methylol group of α D-glucose around the C_5-C_6 axis.

      The γ process of cellulose has been studied by NMR by Mikhailov
et al. (6). The change of spin-lattice relaxation times caused by
deuteration was so small that the mobile unit cannot be only the
primary radical. Thus, the γ process of cellulose has been
attributed to the relaxation of the methylol group as a whole.

The γ process of amylose has been investigated by Nishinari and Fukada by dielectric relaxation (8). The relaxation strength for the rotation of methylol groups around the C_5-C_6 axis has been calculated by using the two site transition theory: a good agreement with the experimental values has been found.

A comparative study of cellophane, amylose and dextran has been performed by Bradley and Carr (2) by dynamic mechanical relaxation. The γ transition was not found in dextran where the methylol groups are missing so the γ mode of amylose and cellophane has been attributed to the rotation of the methylol groups.

From data published in the literature, several molecular processes might be responsible for the fine structure of the γ TSC peaks observed in pullulan: rotations around the α-1.4 or α-1.6 linkages and motions of methylol groups. Further thermally stimulated experiments now in progress on dextran would probably allow a better identification of the various components isolated in the γ mode of pullulan.

ABSTRACT

Thermally stimulated current (TSC) experiments have been performed on pullulan films in order to clarify the relation between the chemical structure and dielectric properties. These complex modes have been experimentally resolved into "elementary" processes characterized by relaxation times obeying Arrhenius equations: the corresponding activation energies, E_i, have been measured. For the peak at $-50°C$: $0.63 < E_i < 1.1$ eV. This mode has been attributed to motions of water molecules introduced during the hydration of the sample. For the peak at $-50°C$: 0.45 eV $< E_i < 0.51$ eV. This mode which is also strongly dependent upon the water content of the sample, may be related to bound water. In slightly hydrated samples, the magnitude of the peak observed at $-130°C$ is not affected by moisture. Regarding the values of activation energies - 0.16 eV to 0.35 eV - two assumptions on the molecular movements responsible for this process, will be discussed: the rotation of methylol groups and the rotation of glucosidic residue around the α-1.4 and α-1.6 linkages.

REFERENCES

1. B.J. Catley, Pullulan Synthesis by aureobasidium Pullulans, in "Microbial Polysaccharides and Polysaccharides", Ed. R.C.Q. Berkeley, G.W. Gooday and D.C. Ellwood, Academic Press, London (1979).
2. S.A. Bradley and S.H. Carr, J. Polym. Sci. Phys. Ed., 14, 111 (1976).

3. J.E. Algie, Colloid and Polymer Sci., 257, 117 (1979).
4. M. Norimoto, Wood Research, 59/60, 106 (1976).
5. A. Sawatari, International Workshop on "Electric Charges in Dielectrics", Kyoto (Japan), Oct. 9-12 (1978).
6. G.P. Mikailov, A.I. Artyukhov and V.A. Sheveley, Vysokomol. Soyed., A11, 553 (1969).
7. S. Yano, H. Hatakeyama and T. Hatakeyama, J. Appl. Polym. Sci., 20, 3221 (1976).
8. K. Nishinari and E. Fukada, J. Polym. Sci., Phys. Ed., 18, 1609 (1980).
9. J. Guillet, G. Seytre, D. Chatain, C. Lacabanne, and J.C. Monpagens, J. Polym. Sci., Phys. Ed., 15, 541 (1977).
10. S. Nomura, A. Hiltner, J.B. Lando, and E. Baer, Biopolymers, 16, 231 (1977).
11. E. Marchal and C. Lacabanne, "Digest of Literature on Dielectrics", A Yelon and M.R. Wertheimer, eds., NAS41, 437 (1977).
12. T. Furukawa, Y. Uematsu, D. Asakawa, and Y. Wada, J. Appl. Polym. Sci., 12, 2675 (1968).
13. T. Takamatsu and E. Fukada, Polymer J., 1, 101 (1970).
14. T. Nedetzka, M. Reichle, A. Mayer, and H. Vogel, J. Phys. Chem., 74, 2652 (1972).
15. M.M. Perlman, J. Appl. Phys., 42, 2645 (1971).
16. Y. Asano, Japanese J. Appl. Phys., 11, 6 (1972).
17. J. H. Ranicar and R.J. Fleming, J. Polym. Sci., Phys. Ed., 10, 1979 (1972).
18. E. Sacher, J. Macromol. Sci. Phys., 11, 6 (1972).
19. R.J. Gable, N.V. Vijayaraghavan, and R.A. Wallace, J. Polym. Sci., Phys. Ed., 11, 2387 (1973).
20. D. Chatain, C. Lacabanne, M. Maitrot, G. Seytre, and J.F. May, Phys. Stat. Sol. (a), 16, 225 (1973).
21. P.K.C. Pillai, K. Jain, and V.K. Jain, Phys. Stat. Sol. (a), 17, 221 (1973).
22. A.E. Blake, A. Charlesby and K.J. Randle, J. Phys. D-Appl. Phys., 7, 759 (1974).
23. J. Van Turnhout, "Thermally Stimulated Discharge of Polymer Electrets", Elsevier, Amsterdam (1975).
24. G. Pfister, W.M. Prest, D.J. Luca, and M. Abkowitz, Appl. Phys. Letters, 27, 486 (1975).
25. Ch. Ponevski and Ch. Solunov. J. Polym. Sci., Phys. Ed., 13, 1467 (1975).
26. T. Hino, J. Appl. Phys., 46, 1956 (1975).
27. S.I. Stupp and S.H. Carr, J. Appl. Phys., 46, 4120 (1975).
28. T. Hashimoto, M. Shiraki, and T. Sakai, J. Polym. Sci., Phys. Ed., 13, 2401 (1975).
29. B. Gross, G.M. Sessler, and J.E. West, J. Appl. Phys., 47, 968 (1976).
30. P. Fischer and P. Röhl, J. Polym. Sci., Phys. Ed., 14, 543 (1976).
31. S. Ikeda and K. Matsuda, Japanese J. Appl. Phys., 15, 963 (1976).

32. I. Chen, J. Appl. Phys., 47, 2988 (1976).
33. P. Alexandrovich, F.E.K. Karasz, and W.J. Macknight, J. Appl.
     Phys., 47, 4251 (1976).
34. M.K. Kenyo, Japanese J. Appl. Phys., 15, 2457 (1976).
35. M. Kryszewski, M. Zielinski, and S. Sapieha, Brit. Polym. J.,
     17, 212 (1976).
36. A. Linkens, J. Vanderschueren, S. Hanchor, and J. Gasiot,
     Europ. Polym. J., 12, 137 (1976).
37. M. Latour and P.V. Murphy, International Workshop on "Thermally
     Stimulated Processes in Solids", Montpellier, France, June
     22-25 (1976).
38. M.R. Grinter and C. Bowlt, J. Phys. D.-Appl. Phys., 9, L61
     (1976).
39. A. Callens, R. de Batist, and L. Eersels, Il Nuovo Cimento,
     33B, 434 (1976).
40. E.J. Sharp and L.E. Garn, Appl. Phys. Letters, 29, 480 (1976).
41. Y. Aoki and J.O. Brittain, J. Polym. Sci., Phys. Ed., 14, 1297
     (1976).
42. T.A. Terziiski, C.R. Acad. Bulgare Sci., 29, 110 (1976).
43. P. Hedvig, "Dielectric Spectroscopy of Polymers", Adam Hilger
     Ltd., Bristol (1977).
44. J. Conway, M.W. Harper, and B. Thomas, J. Phys. D.-Appl. Phys.,
     10, 1131 (1977).
45. S. Celaschi and S. Mascarenhas, Biophysical J., 20, 273 (1977).
46. K. Shindo, Rep. Prog. Polym. Phys. Japan, XX, 369 (1977).
47. T. Mizutani, Y. Suzuoki, and M. Ieda, J. Appl. Phys., 48, 2408
     (1977).
48. T. Tanaka, S. Hayashi, and K. Shibayama, J. Appl. Phys., 48,
     3478 (1977).
49. K. Ikezaki, M. Hattori, and Y. Arimoto, Japan. J. Appl. Phys.,
     16, 863 (1977).
50. J.B. Woodard, J. Elec. Mat., 6, 145 (1977).
51. E. Marchal, H. Benoit and O. Vogl, J. Polym. Sci., Phys. Ed.,
     16, 949 (1978).
52. N. Fukushi, International Workshop on "Electric Charges in
     Dielectrics", Kyoto (Japan), October 9-12 (1978).
53. T. Hashimoto, T. Sakai, and S. Miyata, J. Polym. Sci., Phys.
     Ed., 16, 1965 (1978).
54. S. Radhakrishna and S. Haridoss, J. Appl. Phys., 49, 301 (1978).
55. H.J. Wintle and J. Tuno, J. Appl. Phys., 50, 7128 (1979).
56. K. Nishinari, T. Elsayed, D. Chatain, and C. Lacabanne,
     "Rheology: Applications", Vol. 3, G. Astarita, G. Marucci,
     and L. Nicolais, eds., Plenum Press, New York, p. 533 (1980).
57. Ph. Berticat, private communication, Lyon (1980).
58. K. Nishinari, H. Horiuchi, E. Fukada, Rep. Prog. Polym. Phys.
     Japan, 23, 759 (1980).
59. C. Bucci, R. Fieschi, and G. Guidi, Phys. Rev., 148, 816 (1966).
60. C. Lacabanne, D. Chatain, J. Guillet, G. Seytre, and J.F. May,
     J. Polym. Sci., Phys. Ed., 13, 445 (1975).
61. K. Nishinari, D. Chatain, and C. Lacabanne, ms. in preparation.

62. V.S.R. Rao and J.F. Foster, J. Phys. Chem., 67, 951 (1963).

63. G. Champetier, "Chimie Macromoleculaire", Vol. 2, Hermann (1972).

64. K.M. Grushetskii, Zh. Strukt. Khim., 9, 860 (1968).

65. S. Pérez, M. Roux, J.F. Revol, and R.H. Marchessault, J. Mol. Biol., 129, 113 (1979).

66. I. Tvaroska, S. Pérez, and R.H. Marchessault, Carbohydrate Res., 61, 97 (1978).

67. R.H. Marchessault and S. Pérez, Biopolymers, 18, 2639 (1979).

68. R.G. Zhbankov, J. Polym. Sci., C16, 4629 (1969).